Understanding Environmental Change

Understanding Environmental Change

Rakesh Kumar

RANDOM

RANDOM PUBLICATIONS

NEW DELHI (INDIA)

Understanding Environmental Change

ISBN 978-93-5111-499-4

Published in 2015 in India by

RANDOM PUBLICATIONS

Reprinted 2021

4376-A/4B, Gali Murari Lal, Ansari Road
New Delhi-110 002
Phone : +9111-43580356, 011-23289044, 011-43142548
e-mail: sales@randompublications.com,
info@randompublications.com, randomexports@gmail.com

Type Setting by : Friends Media, Delhi-110089
Digitally Printed at : Replika Press Pvt. Ltd.

Preface

The earth has entered a period of hydrological, climatological, and biological change that differs from previous episodes of global change in the extent to which it is human in origin. To explain or predict the course of the present global environmental changes, one must therefore understand the human sources, consequences, and responses, some of which can alter the course of global change.

To understand global environmental change, it is necessary to focus on the interactions of environmental systems, including the atmosphere, the biosphere, the geosphere, and the hydrosphere, and human systems, including economic, political, cultural, and sociotechnical systems. Human systems and environmental systems meet in two places: where human actions proximately cause environmental change, that is, where they directly alter aspects of the environment, and where environmental changes directly affect what humans value.

Almost all human activity has some potential relevance to global change. Researchers in a number of fields have studied human-environment interactions, usually within the boundaries of single disciplines and almost always below the global level. They have demonstrated that a complex of social, political, economic, technological, and cultural variables, sometimes referred to as driving forces, influences the human activities that proximately cause global change.

This book provides a basic grounding in the science of global environmental change. The philosophy of the book is based on the view that an understanding of natural global change is necessary to assess the impact of human activity. The sheer excitement of the science, the wealth of facts and the clarity of the exposition will make it an ideal undergraduate and masters degree course book.

Author

Contents

1

Environmental Systems and Change

The biophysical environment is the biotic and abiotic surrounding of an organism, or population, and includes particularly the factors that have an influence in their survival, development and evolution. The naked term environment can make reference to different concepts, but it is often used as a short form for the biophysical environment. This practice is common, for instance, among governments, that usually name their departments and agencies dealing with the biophysical environment with denominations like Environment Agency. Whereas the expression "the environment" is often used to refer to the global environment, usually as referred to humanity, the number of biophysical environments is countless, given that it is always possible to consider an additional living organism that has its own environment.

The biophysical environment can vary in scale from microscopic to global in extent. They can also be subdivided according to their attributes. Some examples may be the marine environment, the atmospheric environment and the terrestrial environment.

Life has to be adapted to its environment conditions. Temperature, light, humidity, soil nutrients, etc., all this has an influence in the species that can inhabit a particular environment. But life is not passive, and its activity modifies in various forms the environment conditions. Some long term modifications along the history of our planet have been dramatic, like the incorporation of oxygen to the atmosphere. The process consisted in the break down of carbon dioxide by anaerobic microorganisms that used the carbon in their metabolism, and released the oxygen to the atmosphere, and thanks to this, plant and animal life, that need oxygen, could emerge (Great oxygenation event). Other interactions are more immediate and simple, like the effect that forest trees have in the smoothing of the temperature cycle, compared to non protected neighboring areas.

Environmental change is a significant and lasting change in the statistical distribution of weather patterns over periods ranging from decades to millions of years. It may be a change in average weather conditions, or in the distribution of weather around the average conditions (i.e., more or fewer extreme weather events). Environmental change is caused by factors that include oceanic processes (such as oceanic circulation), variations in solar radiation received by Earth, plate tectonics and volcanic eruptions, and human-induced alterations of the natural world; these latter effects are currently causing global warming, and "climate change" is often used to describe human-specific impacts.

Scientists actively work to understand past and future climate by using observations and theoretical models. Borehole temperature profiles, ice cores, floral and faunal records, glacial and periglacial processes, stable isotope and other sediment analyses, and sea level records serve to provide a climate record that spans the geologic past. More recent data are provided by the instrumental record. Physically based general circulation models are often used in theoretical approaches to match past climate data, make future projections, and link causes and effects in climate change.

Man has only been evident on our planet for a few millions of years. None the less, the environmental changes that have taken place during the time we have inhabited the Earth have not been slight. Changes in climate, sea-levels, vegetation belts, animal populations, soils and landforms, have been both many and massive. Some changes are still taking place, and have been of such an order as to influence markedly humans and the landscape. In the last 20 000 years alone - and humans, as already noted, have been in existence one hundred times longer than that - the area covered by glaciers has been reduced to one-third of what it was at the glacial maximum; the waters thereby released have raised ocean levels by over one hundred metres; the land, unburdened from the weight of overlying ice, has locally risen by several hundred metres; vegetation belts have swung through the equivalent of tens of degrees of latitude; permanently frozen ground has retreated from extensive areas of Europe; the rainforest has expanded; desert sand fields have advanced and retreated; inland lakes have flooded and shrunk; and many of the finest mammals have perished in the catastrophe called 'Pleistocene overkill'. Even at the present time, smaller fluctuations of climate are leading to changes in fish distributions in northern waters, marked fluctuations in valley glaciers, extensive flooding or desiccation of lakes, and difficulties for agricultural schemes in marginal areas.

During the past few million years humans have gradually diffused over the Earth and have now become powerful agents of environmental change in their own right. The oldest records of human activity - crude stone tools which consist of a pebble with one end chipped into a rough cutting edge - have been found in conjunction with bone remains from various parts of East Africa.

Subdividing Geological Time

The terminology used to describe some of the natural events and changes which provide the backcloth of human history needs to be discussed here. The major geological divisions traditionally employed are as follows: the most recent *era* is called the Cainozoic ('recent life') and this has been divided into two *periods:* the Tertiary and the Quaternary. The use of these two terms in a geological sense dates back to the 1760 classification scheme of the Italian geologist Giovanni Arduino in which he discussed geological time in terms of four main phases - primary, secondary, tertiary, and quaternary. This terminology was further established by Desnoyers in his 1829 study of Tertiary and Quaternary marine deposits in the Paris basin. As geological understanding increased through the nineteenth century, this basic four-part scheme was modified and elaborated; yet, while the terms 'primary' and 'secondary' dropped out of use, the terms 'tertiary' and 'quaternary' survived as the names of the two most recent geological periods.

These periods are in turn divided into various *series* or *epochs,* with the Pliocene, for example, being the final series of the Tertiary. The Pleistocene and Holocene series together constitute the Quaternary Period - a mere 0.04 per cent of the 4600 million years since the formation of the Earth. Using the analogy of Earth's history being represented by a single day, this is equivalent to the last half-minute of that day. Yet, as already mentioned, within this brief period there have been major climatic and environmental changes which have greatly influenced the world we inhabit, both in terms of the physical landscape and the evolution and distribution of organisms. The vast majority of Quaternary time is spanned by the Pleistocene ('most recent'), but a sliver of very recent geological time which extends from about 11 500 years ago to the present is given status as a separate series, termed the Holocene ('wholly recent'), because of the development of agriculture and civilization during this time span.

It is difficult to draw boundaries between some of these units for they represent terminological and classificatory devices rather than reality. As a consequence, there is dispute among authorities about how the Cainozoic should be subdivided. Recently the International Commission on Stratigraphy (ICS) has established usage of the terms *Palaeogene* and *Neogene* to distinguish the early Cainozoic (prior to 22.5 million years) from the later Cainozoic extending to the present. There has been debate recently over the formal status of the Quaternary in the Geological Time Scale because of the ICS proposal to redefine the Quaternary as a sub-era within a 'Neogene Period'; and some authorities have even advocated the abandonment of the term Quaternary in favour of the term 'Late Neogene', arguing that the Quaternary only represents an intensification of climatic trends associated with the onset of the Neogene. Similarly, because the climatic transition between the Pleistocene and the Holocene is no different in character from the transitions between the various preceding glacials and interglacials, some authorities would consider the Holocene as the most recent stage of the Pleistocene rather than as a separate series.

Whatever terms we may use, it is still clear that changes in environment have occurred over a wide range of timescales which range from the minor fluctuations within the period of instrumental record (with durations of the order of a few years or decades) to the major geological periods with durations of many millions of years. The shorter-term changes include such events as the period of warming that took place in the first decades of the twentieth century. The changes with durations of hundreds of years were characteristic of the Holocene and include various phases of glacial advance and retreat such as the Little Ice Age between about AD 1500 and 1850. The fluctuations *within* the Pleistocene, consisting of major glacials and interglacials, lasted for the order of 10 000 to 100 000 years, while the Pleistocene itself consisted of a group of glacial and interglacial events that lasted in total about two million or so years. Such major phases of ice age activity in Earth's geological history appear to have been separated by about 250 million years.

Evolution of Ideas

Appreciation of the fact that the world had such a history of environmental change emerged at much the same time that it was appreciated that the world had a history that was longer than the constricting 6000 years of Archbishop Ussher's biblical time span. This seventeenth-century concept, which saw the Earth as having been created in 4004 BC, was relatively uncontested until the end of the eighteenth century; and it was accompanied by the belief that much of the denudation and deposition evident on the face of the Earth could be explained through the agency of Noah's Flood and other catastrophes. Gradually, however, these ideas were shown to be erroneous. In particular, Hutton and Playfair, the Edinburgh scientists, are often regarded as the men who most effectively propagated the new ideas, for in the geological record they saw 'no vestige of a beginning, no prospect of an end'. They realized that the complexity of the sedimentary record could be explained through the operation of processes akin to those of the present day extending over a long time span. A full discussion of the development of these ideas is provided by Chorley et al.

In spite of this convergence of opinion from numerous sources, these ideas were not easily assimilated into prevailing dogma, and for many years it was still believed that glacial till, called drift, and isolated boulders, called erratics, were the result of marine submergence, much of the debris, it was thought, having been carried on floating icebergs. Charles Lyell noted debris-laden icebergs on a sea-crossing to America, and found that such a source of drift was more in line with his belief in the power of current processes - the uniformitarian belief - than a direct glacial origin. For years even glacial or subglacial depositional features like eskers were thought to be of a marine origin, and were classified into fringe eskers, bar or barrier eskers, and shoal eskers. Moreover, in Britain, some drift deposits contained sea shells, a fact that gave prima facie support to marine ideas.The study of environmental change progressed somewhat further in the 1860s when Archibald Geikie, A.C. Ramsay, and T.F. Jamieson showed that the spread of the glaciers had not been a single event, and that there had been various stages of glaciation, represented by moraines, which had been separated by warmer phases, called interglacials.

Ice Age Discovery

The idea that climate and other aspects of the environment had fluctuated or changed during this enlarged span of time resulted initially from the discovery that European glaciers had formerly extended further than their current limits. In 1787 de Saussure recognized erratic boulders of Alpine rocks on the slopes of the Jura Ranges and Hutton reasoned that such far-travelled boulders must have been glacier-borne to their anomalous positions. Playfair extended these ideas in 1802 but it was in the 1820s that the Glacial Theory, as it came to be known, really became widely postulated. Louis Agassiz, who was one of the originators of the term *Eiszeit* or Ice Age, did more than anyone else to promote the concept of past glaciation at the continental scale.

Lower Latitudes

In areas outside those which were subjected to Pleistocene glaciation, other types of fundamental change were recognized as having taken place, though Louis Agassiz, after a journey to Brazil, postulated somewhat wildly that even equatorial regions had undergone glaciation. He believed that the Amazon Basin had been overwhelmed, but he seems, like certain others at the time, to have allowed his enthusiasm to convince him that what was in reality deeply weathered soil and boulders produced by intense chemical activity under tropical conditions, was glacial boulder clay. A truer appreciation of the effects of climatic change in the tropics and sub-tropics came from Jamieson, Lartet, Russell, and G.K. Gilbert, all of whom studied the fluctuations of Pleistocene lakes in semi-arid areas, as represented by old strandlines and deltas. Their work established the supposed general relation between high-latitude glacials and mid- and low-latitude pluvials, and Russell was able to tie in the moraines of the former Sierra Nevada glaciers with old strandlines around Lake Mono in California. Gilbert demonstrated that Lake Bonneville (Utah), in much the same way as the glaciers, had fluctuated several times with alterations of high water and phases of desiccation. Other geologists investigating the west of the United States also believed that a diminution of rainfall might account for some of the anomalous drainage features they encountered on their travels, and the term 'pluvial' was coined by Alfred Taylor in 1868. It means basically a period of comparatively abundant moisture in regions beyond the limits of the ice-sheets.

Thus in the south-west of the United States we have one of the germs of the idea that low-latitude pluvials were synchronous with high-latitude glacials. A corollary of this view, that post-glacial times will have been characterized by desiccation, spread through many parts of the world. A particularly notable focus in the early years of the twentieth century was Central Asia, particularly after the great explorations of the Tarim Basin, Lop Nor, and Tibet. In 1903 Andrew Carnegie financed an expedition to Turkestan and Persia, of which, among others, Raphael Pumpelly, W.M. Davis and E. Huntington were members. In 1905-6 Huntington was able to go a second time to Asia. On these journeys, he noted the evidence for ruined cities and abandoned settlements, and recognized that lake terraces indicated a 'gradual desiccation of the country from early historical times down to the present'. After visiting the famed Tarim Basin, he wrote his *Pulse of Asia* and moved away from any simple idea of progressive desiccation to a view embracing multiple large fluctuations. He believed that the pulsations of climate had served as a driving force in the history of Eurasia, forcing nomadic invaders to overrun their more civilized neighbours whenever the climatic cycle reached a trough of aridity.

At much the same time as Huntington put forward his views, Prince Kropotkin put forward a simple view of progressive desiccation:

> Recent exploration in central Asia has yielded a considerable body of evidence, all tending to prove that the whole of that wide region is now, and has been since the beginning of the historic period, in a state of rapid desiccation... It must have been the rapid desiccation of this region which compelled its inhabitants... pushing before them the former inhabitants of the lowlands to produce those great migrations and invasions of Europe which took place during the first centuries of our era.

The concept of post-glacial desiccation was much favoured in Indian literature of the 1950s and 1960s. In particular, fears were expressed that the Thar Desert was on the move eastwards. In East Africa, Nilsson proposed a possible correlation between African pluvials and Alpine glaciations, and in his influential work on climatic geomorphology, Büdel maintained that the Sahara had been squeezed during glacial times as the westerly depression tracks were displaced from the north and rainforest and savanna encroached from the south.

Distrust of long-distance correlations gradually arose, and was forcefully voiced by Zeuner:

> the suggested correlation of certain tropical pluvials with certain glaciations of Europe is not only unfounded, but in some cases definitely wrong. It was

> chiefly based on the mistaken, though most widely accepted, assumption of the strict contemporaneity of tropical glaciations and pluvials with European glaciations. This is a typical example of uncritical acceptance of an assumption born from the desire of simplifying, which is so current in scientific thought, though it often violates the facts.

In the 1950s we see the development of an increasing uneasiness about correlations between glacial events in high latitudes and pluvial events in low. This change of attitude was prompted by a combination of new dating technologies and by new types of evidence for environmental reconstruction and is well represented in the work of Zeuner. Of especial importance was the recognition that some tropical areas, far from being highly desiccated at the present time, had experienced far more severe desiccation at some point, or points, in the past. Instrumental in this was the identification of ancient sand seas *(ergs)* in areas that are now relatively moist, most notably on the south side of the Sahara.

Glacial and Interglacial Sequences

After the existence *of Die Eiszeit* had been established in the 1820s and 1830s, it gradually became evident that the Ice Age was a complex event in which there had been a number of glacial advances and retreats. Although monoglaciation had its adherents well into the twentieth century, the general view was of multiple glaciations.

Table 1. Some classical schemes of subdivision in different regions.

Alps	*Northern Europe*	*British Isles*	*North America*
WÜRM	WEICHSEL	NEWER DRIFT	WISCONSIN
R-W	Eemian	Ipswichian	Sangamon
RISS	SAALE	GIPPING	ILLINOIAN
M-R	Holstein	Hoxnian	Yarmouth
MINDEL	ELSTER	LOWESTOFT	KANSAN
G-M	Cromerian	Cromerian	Aftonian
GÜNZ	NEBRASKAN		

This was certainly the case for James Geikie, who reviewed the evidence, particularly in the British Isles. In central Europe, between 1901 and 1909, A. Penck and E. Brückner produced a major work, *Die Alpen im Eiszeitalter.* For decades this formed the basis for much glacial chronology, and in particular they proposed a classic fourfold glaciation model of the

Pleistocene. By studying what was left of old moraines, they were convinced that there had been four great glaciations of different intensities in the Alps. They also proved, from the preservation of plant remains at some sites, that some of the intervening periods were fairly mild. They further demonstrated the correspondence between these four glaciations and the successive gravel terraces of the Rhine and other rivers. They named the four glacial periods after valleys in which evidence of their existence occurred: the Günz, Mindel, Riss, and Würm.

The Holocene and the Late Glacial

'Holocene' was introduced by the French palaeontologist Paul Gervais in 1869 as a term for the last 10 000 years or so of Earth history. At the end of the nineteenth century it was recognized in Norway and Sweden that during the Holocene there had been a series of changes in vegetation. Axel Blytt, a Norwegian, argued that the remains of different plants (plant macrofossils) preserved at different levels within peat deposits indicated a pattern of species immigration into Norway following the last glacial, and that the sequence of these plant remains also reflected past climatic changes. He introduced the terms 'arctic', 'boreal' and 'atlantic' to characterize the different plant assemblages and the inferred climates. Sernander, a Swede, further developed the terminology and applied the concept more widely to characterizing the postglacial vegetation development of Scandinavia, and the 'Blytt-Sernander' scheme was born. As the sequence of peat layers, their degree of decomposition (humification) and associated plant remains appeared fairly uniform over large areas, many workers believed that the climatic subdivisions of the Blytt-Sernander scheme were broadly synchronous across north-west Europe. Hence, before the development of radiocarbon dating, the Blytt-Sernander scheme was widely used for correlating Holocene terrestrial deposits.

Shortly after the development of the Blytt-Sernander scheme, von Post introduced the method of pollen analysis (palynology) as a means of reconstructing past vegetation changes in an area. As the ideas of the Blytt-Sernander scheme were common currency among palaeoecologists in the first half of the twentieth century, pollen analysts (working with lake or peat sediments) generally interpreted their Holocene pollen data within the Blytt-Sernander framework. The first such detailed pollen-based vegetational reconstruction (pollen zonation scheme) was developed by Jessen for

Denmark. Godwin published a similar sequence of Holocene pollen zones for England and Wales, and Jessen applied the system to Ireland. While each of these zonation schemes differed to some extent in plant species composition, they were all based on the assumption that the zonations were broadly synchronous over large parts of north-west Europe, and that the inferred changes in Holocene climate correlated with the Blytt-Sernander subdivisions. For example, pollen zone VIIa was seen to represent the height of the Atlantic period of the Blytt-Sernander scheme when it was inferred that temperature and precipitation were higher than at present and that north-west Europe experienced a 'climatic optimum'.

It is worth noting, however, that the idea of an Atlantic climatic optimum' in Europe contrasted with interpretations for the same time period in America. In the drylands of North America it was, from the human point of view, far from optimal. Evidence for a period of intense drought, with dune reactivation in the High Plains, the drying of lakes and trenching of valley floors led to the adoption of the term 'altithermal' or 'hypsithermal'.

Until the 1970s, most pollen analysts in Europe interpreted their data in relation to the Blytt-Sernander and Jessen-Godwin frameworks, and sometimes pollen-based vegetation reconstructions were thought to be flawed if they could not be fitted within this system. With the advent of radiocarbon dating, and its widespread application to terrestrial deposits, particularly in the 1970s, it has been shown that the palaeoecological and climatic subdivisions of the Blytt-Sernander scheme do not represent contemporaneous changes across northwest Europe, and it is now known that the pattern of Holocene climatic change is much more complicated than the scheme suggests - both temporally and spatially.

Geomorphologists came to realize that, notwithstanding the views of people like Raikes, who argued that the climate of the Holocene was rather stable, the climatic changes inferred from the study of plant remains and pollen had considerable geo-morphological significance for fluvial systems and, above all, for glaciers. Moss and Nelson used the term 'neoglaciation', which was subsequently defined by Porter and Denton as a 'cool geologic-climate unit... indicating a probably world wide synchrony of glacier fluctuations in response to climatic change'. Radiocarbon dating revealed that glaciers had experienced a series of advances and retreats during the Holocene.

Around the time that the Holocene climate-vegetation subdivisions were being established, evidence was also found pointing to a major climatic deterioration that had taken place near the transition between the Pleistocene and the Holocene. Towards the end of the last glacial phase (a period of time known as the Late Glacial when the climate was generally warming and ice-sheets were receding) there was a short, but intense, cold phase known as the Younger Dryas, when glacial conditions were reestablished in many parts of the world. For example, glaciers returned to the Scottish Highlands and intense valley excavation occurred in southern England.

The Younger Dryas was first recognized in 1901, at Allerød in Denmark, where it was discovered that lake sediments showed a pattern of organic mud with birch remains sandwiched by lower and upper clay layers containing the remains of cold-adapted plants. This suggested that the warming following the end of the last glacial was interrupted by a shift back to cold conditions before the full interglacial conditions of the Holocene became established. Pollen analysis of sediments laid down immediately after full glaciation in other parts of north-west Europe showed a similar 'tripartite' sequence with upward progression from arctic/alpine pollen types, to dwarf-shrub and birch pollen, and back to arctic/alpine types.

Sea-level Change

One of the great concerns of Quaternary scientists in the first half of the twentieth century were worldwide (eustatic) changes in sea-level. Suess, the Austrian scholar, in volume 2 of *Das Antlitz der Erde,* believed that evidences of continental transgressions and regressions pointed to a remarkable synchroneity of swings of sea-level in widely spaced areas of the world. He later maintained that many continental areas had been absolutely stable during geological time, but that there had also been rapid subsidence of the ocean bottoms, which caused sea-levels to fall in a punctuated manner, creating staircases of terraces.

Geomorphologists, finding flights of marine terraces, sought to correlate them over wide areas on the basis of altitude differences. It was argued that strandlines occurred at similar heights around much of the Mediterranean. At a high level was the Calabrian (Pliocene or Pleistocene) bench with various others at lower levels. Baulig was an exponent of the idea that erosion surfaces in the Massif Central were the result of eustatic changes, and influenced Wooldridge, who argued that a supposed bench in southeast England was a marine terrace of Calabrian age. There was also a view that a progressive decline in sea-levels had taken place during the Pleistocene, with terraces forming in the main interglacials and with progressively lower low sea-level stands during the four main glacials.

Perhaps one of the firmest advocates of eustatic sea-level correlations was Zeuner who produced a table, using the Mediterranean terminology, which purported to show the correspondence of shorelines by altitude in Algeria, the south of France, the Channel Islands, northern France, southern England, South Africa, South Australia, the Sunda Archipelago and North America! He produced a table of sea-levels and approximate dates which he hoped would provide an absolute chronology of the Pleistocene that could be extended over the whole Earth.

Such views were criticized, not least because of the large degree of tectonic instability found in an area like the Mediterranean Basin, but also because there was limited valid chronological control on the ages of the terraces. The eustatic theory lingered on tenaciously, but was plainly a great over-simplification of a complex problem.

However, the debates over global eustatic correlations should not deflect us from appreciating the significance of glacio-eustatic changes during the Pleistocene and Holocene. As Daly pointed out, glacial-age sea-levels stood at around 100 m below the sea-level of today and had huge implications for the evolution of such features as coral reefs, deltas and river valleys. Moreover, by the mid-1960s, radiocarbon dating had given a picture of the nature of sea-level rise in post-glacial times, though there was some debate about whether Late Holocene sea-level positions oscillated to points both above and below the present level.

Ideas of glacio-isostasy, whereby the addition and removal of ice-caps caused crustal deformation, were developed initially in Scotland by Jamieson. They were rapidly taken up in North America where the term 'isostasy' was introduced by Dutton. Gilbert extended the idea to include the effects of water loading associated with the expansion of pluvial Lake Bonneville - hydro-isostasy. By the close of the 1880s, therefore, the hypotheses of glacio-isostasy and hydro-isostasy had been outlined. De Geer produced maps of coastal deformation for Scandinavia, and by the beginning of the twentieth century 'the pattern of crustal response to glacial unloading in Scandinavia was being reconstructed in impressive detail, providing a model for studies in other glacio-isostatically affected areas'. Substantial work was also undertaken on glacio-isostatic warping of the Great Lakes of North America. The deformation of shorelines in Scotland was discussed by Wright and was the subject of intensive research by Sissons and his pupils.

In areas that were glaciated the history of sea-level change is usually the history of the race or struggle between glacio-eustatic and glacio-isostatic effects. Wright argued that this was responsible for the complex form of raised beaches in areas like Scotland and the Baltic.

Early Climate History

Little is known about the very early climatic history of the Earth, for rocks deposited before about 2800 million years ago are generally so metamorphosed that it is extremely difficult to extract climatic information from them. In the early Proterozoic evidence from Canada suggests that the first known glaciation occurred. It was probably a complex event with recurring glacial events rather than a continuous glaciation, and it was followed by a relatively warm phase that persisted to around 950 million years ago. During the late Precambrian at least three major glacial episodes occurred, each lasting about 100 million years and centring on about 615 million, 770 million, and 940 million years ago.

Toward the end of the Proterozoic, the record of climate improves because the sedimentary record is better preserved and animals with shells that could be preserved as fossils had spread through the oceans. In general climatic conditions between 570 million and 225 million years ago were relatively warm, though in the Sahara there is evidence of a brief glacial event in the late Ordovician around 430 million years ago. The Permocarboniferous was characterized by a long glaciation from 330 to 250 million years ago, and ice-sheets were extensive on the supercontinent of Pangaea. During the succeeding Mesozoic era the evidence of deep-sea cores becomes of greater utility. Temperatures ranged from 10-20 °C at the poles to 25-30 °C at the equator. There were, however, fluctuations in temperatures, with slight cooling taking place at the start of the Jurassic and with marked high-latitude warming taking place during the first half of the Cretaceous. Global cooling took place towards the end of the Cretaceous but there was a significant increase in temperatures at the onset of the Tertiary. However, a long-term cooling trend commenced shortly after the start of the Eocene epoch some 50 million years ago, following a major spike of warmth at the Palaeocene/Eocene boundary. The Cainozoic suffered various marked short-term cooling episodes: particularly in the mid-Palaeocene (*c.* 58 million years ago), in the Middle Eocene (*c.* 45 million years ago), and at the Eocene-Oligocene boundary (*c.* 34 million years ago). As described in more detail below, there is some evidence that the Antarctic glaciers may have come into existence in the Eocene and that about 10-15 million years ago mountain glaciers came into existence in the Northern Hemisphere.

Cainozoic Climatic Decline

Although global temperatures varied considerably through the Tertiary Period, on average the globe was significantly warmer during the Tertiary than during the Quaternary. The average global temperature of the Tertiary reached a peak during the Eocene between 52 and 50 million years ago with mean annual temperatures in many places over 10 °C warmer than at present. In the mid-Eocene there was no ice at the Poles, tropical climates extended 10 to 15° higher latitude than today and subtropical plants survived as far as 60° north. Remarkably, fossils of alligators and other warmth-loving animals dating from this time have been found as far north as Ellesmere Island to the west of Greenland. With no polar ice, sea-level was consequently much higher than today, and the atmosphere was more humid. By the late Eocene (38 million years ago) the climate had cooled enough for the Antarctic ice-sheet to begin its formation, and temperatures dropped relatively sharply at the Eocene/ Oligocene transition (34 million years ago). Despite later warming trends, temperatures never came close to reaching the peak attained during the Eocene, and by the end of the Tertiary (in the late Pliocene Epoch about 3 million years ago) ice-sheets were also building up in the high northern latitudes.

Given the high degree of climatic variability during the Tertiary, it may be over-simplistic to talk of 'average Tertiary conditions' for comparison with the Quaternary. However, many workers have attempted to estimate the overall decline in temperatures through the entire Tertiary. On the basis of deep-sea sediment core studies, Emiliani suggested that there was a broad temperature decrease amounting to about 8 °C for the middle latitudes during the Tertiary. A similar temperature depression is indicated in the Equatorial Pacific, but on the basis of land flora studies, somewhat greater changes than these have been proposed for the western United States between 40°

and 50°N, and the picture for other areas is the same. The pattern of temperature decline for Pacific Ocean water, with a progressive shift in isotherms as the climate became cooler.

In Australia a comparable sequence of decline of Tertiary temperatures has been proposed, based on the former great extent of the plants *Araucaria* and *Agathis* into Tasmania. At the present time they are limited to Queensland and the warmer parts of Australia. However, rainfall decline may have been equally important in such changes of floral distributions, and Gentilli has said that 'areas that now receive 12.5 cm of rain a year must then have received at least 125 cm with no rainy season. If there was even a short dry season the annual rainfall must have reached some 200 or 250 cm for these laurisil-vae to grow as they did.' Large trees then existed in the Lake Eyre Basin and other stretches of the 'dead heart' of Australia.

Environments of the Tertiary

The North Atlantic region in the early Tertiary may have been characterized by a widespread, tropical moist forest type. Indeed, the classic interpretation of the fossil plant evidence from the Palaeocene London Clay of southeast England, made by Reid and Chandler, was that it represented a flora comparable to that of the Indo-Malayan region of today. Whether there really was a tropical rainforest environment in southern Britain in the Eocene is, however, a matter for debate. As Daley has pointed out, 'Such a climate, with its markedly uniform character throughout the year, could not have existed at the latitude of southern Britain (40°N) during the Eocene, since seasonal climatic variations would undoubtedly have occurred.'

Nevertheless the picture of early and middle Eocene warmth is confirmed by many studies. For example, Collinson and Hooker report that beds of that age contain the highest proportion of potential tropical flora (up to 82 per cent) and the smallest proportion of potential temperate nearest living relatives (up to 42 per cent). Collinson and Hooker. consider that: 'A great many of the near living relatives of fossils co-exist today in paratropical rainforest in eastern Asia. Many are large forest trees, others lianas.

The greatest diversity of these elements in the Early and Middle Eocene suggests the presence of a dense forest vegetation.' Tree rings were also particularly large, indicating rapid growth in response to high temperatures.

The climatic deterioration at the end of the Eocene, which appears abrupt or gentle, according to the type of evidence used, created a different climatic regime in the Oligocene. Then the climate of Britain was more comparable to that of a region like the south-eastern United States, though there may have been very dry intervals, as indicated by the presence of gypsum evap-orite crusts and ancient desert dunes in the Ludien and Stampian sediments from the Paris Basin. However, by Pliocene times, the 'tropical' vegetation of the North Atlantic region was being replaced by a largely deciduous warm temperate flora. The last appearance of palms north of the Alps in Europe is in the Miocene flora at Lake Constance.

Immediately before the Ice Age, in a period which the Dutch call Reuverian B, the climate of the present temperate latitudes of the Northern Hemisphere seems to have favoured the development of woodland. At that time varied woodlands extended from the Atlantic coasts to the Sea of Japan, a picture that has not been repeated since. Over wide reaches of the present warm and cool temperate latitudes, temperature and rainfall conditions were particularly favourable, and resembled those of a subtropical climate. In central and eastern Europe, temperatures were probably 3-5 °C higher than they are now, and precipitation levels were several hundred millimetres higher.

Explaining Tertiary Climates

The causes of a warm Palaeogene in Britain are both local and global. At a local scale, Britain was at a lower latitude than today, being 10-12° further south in the Palaeocene. Movements of the continents may also have affected global climates by creating a unique configuration of the continental masses. Parrish has suggested that:

> By the beginning of the late Cretaceous, global palaeo-geography had entered a mode of maximum continental dispersion, and the monsoonal circulation that had dominated palaeoclimatic history during the previous 250 million years had disappeared. From this time until the collision of India and Asia re-established strong monsoonal circulation during the Miocene, continental palaeoclimates were characterized by the zonal circulation that is the fundamental component of atmospheric circulation on Earth.

Global Palaeogene warmth is thought to have been caused largely by elevated levels of greenhouse gases in the atmosphere. Carbon dioxide levels were high because of frequent volcanic eruptions associated with high rates of sea floor spreading and plate subduction. The warming spike at around 52 million years ago is thought to have been caused by the rapid release of large quantities of methane into the atmosphere which further intensified the greenhouse effect (methane, molecule for molecule, is over 20 times more effective as a greenhouse gas than carbon dioxide). The source was the breakdown of some 1200 gigatonnes of gas hydrates - a combination of water and methane normally kept solid at high pressures and low temperatures beneath the oceans.

The causes of the 'Cainozoic climate decline' are still not fully understood, but the trend seems to be associated with the break-up of the ancient supercontinent of Pangaea into the individual continents we know today. Round about 50 million years ago Antarctica separated from Australia and gradually shifted southwards into its present position centred over the South Pole. At the same time Eurasia and North America moved towards the North Pole. As more and more land became concentrated in high latitudes ice-caps could develop, surface reflectivity increased, and as a consequence climate probably cooled all over the world.

Late Cainozoic cooling may have promoted the expansion and development of some of the world's great deserts. For instance, in late Cainozoic times (van Zinderen Bakker 1984) aridity became a prominent feature of the Saharan environment, probably because of the occurrence of several independent but roughly synchronous geological events:

(i) As the African plate moved northwards there was a migration of northern Africa from wet equatorial latitudes (where the Sahara had been at the end of the Jurassic) into drier subtropical latitudes.

(ii) During the late Tertiary and Quaternary, uplift of the Tibetan plateau had a dramatic effect on world climates, helping to create the easterly jet stream which now brings dry subsiding air to the Ethiopian and Somali deserts.

(iii) The progressive build-up of polar ice-caps during the Cainozoic climatic decline created a steeper temperature gradient between the equator and the poles, and this in turn led to an increase in trade wind velocities and their ability to mobilize sand into dunes.

(iv) Cooling of the ocean surface may have reduced the amount of evaporation and convection in low latitudes, thus reducing the amount of tropical and subtropical precipitation.

The Tertiary/Quaternary Transition

The Quaternary has long been distinguished from the preceding Tertiary by global cooling associated with the expansion of glaciers, particularly across high to mid-latitude land masses in the Northern Hemisphere. However, the precise time of transition from the Tertiary to the Quaternary (which equates with the Pliocene/Pleistocene series boundary) has been, and continues to be, a matter of interpretation and debate. It was originally thought, from the traditional Alpine glaciation model of Penck and Brückner, that the Pleistocene began about 600 000 years ago; however, over succeeding decades it became clear that the Pleistocene needed a much longer chronology than that.

Furthermore, the traditional view of glaciation being confined to the Quaternary must be discarded. Previously it had been widely believed that the period from the Triassic through the Tertiary was a lengthy span when ice-sheets and glaciers were absent, and when climatic fluctuations were less frequent and less severe than they were to be in the Pleistocene. However, this idea became increasingly untenable as more was learnt about Tertiary environments, particularly from studies of ocean sediments. For example by studying the *Globigerina* type of Foraminifera preserved in ocean-sediment cores, Bandy found evidence for a major expansion of polar faunas no less than between 10 and 11 million years ago. This expansion was followed, he suggested, by another mid-Pliocene expansion between 5 to 7 million years ago, and the classic Pleistocene expansion some 3 million years ago. Bandy also argued that the magnitude of planktonic faunal changes in the Tertiary 'are almost as great as those of the classic Quaternary'. As early as the 1970s Flint remarked that 'perhaps the most stirring impression produced by recent great advances in the study of the Quaternary period is that the Quaternary itself is losing its classical identity.

Antarctic glaciation stretches back into the late Eocene, perhaps as much as 38 million years ago. The appearance of iceberg-rafted debris in Southern Ocean deep-sea cores suggests that the east Antarctic glaciers reached a full-bodied stage somewhat before 5 million years ago. Another core from the area led to the even more striking conclusion that glacial conditions were present in the Eocene, since a coexistence of micro-fossils of Eocene age, and of sediments of glacial type, has been proved. In eastern Antarctica ice-field initiation dates back to the lower Miocene. Confirmatory evidence of Eocene glaciation in Antarctica is provided by the nature of dated volcanic materials on that continent. Volcanoes that have erupted beneath an ice-sheet display a suite of textural and structural characteristics that are especially distinctive in volcanoes composed of basaltic lava, and such materials have been found in Antarctica back to the Eocene. Similar evidence implies also that glaciation may well have been uninterrupted up to the present in this South Polar region.

Given the variable nature of Tertiary climates and the early onset of Cainozoic cooling, it is difficult to make any division which is both logical and rigid between the Pleistocene and the Pliocene. Indeed the Villafranchian (the earliest unit of the European Pleistocene), together with its marine equivalent, the Calabrian, have only been assigned to the Pleistocene rather than to the Pliocene since 1948.

On faunal grounds the Pleistocene has come to be regarded as the time when many of the modern genera, including elephant, camel, horse, and wild cattle, first appeared. Attempts have also been made to place the boundary between the Upper Pliocene and the Villafranchian of the Lower Pleistocene by using tectonic breaks in the stratigraphic succession, though this provides a generally inadequate and unusable correlation basis, except on a very local scale. In Britain the division between the Pliocene and the Pleistocene is placed at the boundary between the Coralline Crag and the Red Crag of East Anglia. At this point there is a relatively clear stratigraphic break, a marked increase in the proportion of modern forms of marine Mollusca, and of Mollusca of northern aspect, and the first arrival (in the Red Crag) of elephant and horse.

In Europe the base of the Pleistocene has been identified by the appearance in Late Cainozoic sediments occurring in various parts of Italy of a cold-water marine fauna (the Calabrian) differing from the underlying Pliocene fauna. The newer fauna are characterized by the appearance of a dozen species of North Atlantic molluscs, and by certain Foraminifera. Thus, the Olduvai geomagnetic event has been used as a marker to identify the beginning of the Pleistocene series throughout the world. At the time that the Vrica Global Stratotype Section and Point (GSSP) was established, it was thought that the point of transition in the section dated to 1.64 million years ago; but since then, improvements in the palaeomagnetic timescale have pushed the date back to 1.81 million years ago. Those who are in agreement with this way of marking the boundary can be said to favour the 'short Quaternary chronology'.

However, many other researchers favour a longer Quaternary in which the Pliocene/Pleistocene boundary is placed at around 2.6 million years ago - the time when evidence in deep-sea sediment cores from the North Atlantic and North Pacific Oceans suggests that substantial glaciation was underway. Although there have been mountain glaciers in the Northern Hemisphere for at least 10 million years (this is shown, for instance, in the White River Valley area of Alaska where lithified glacial deposits are inter-bedded with volcanic lavas dated to between 9 and 10 million years old), ice-sheets only began forming between about 3 and 2.5 million years ago. Through a recent study of deep-sea sediments from the subarctic Pacific Ocean Haug et al. suggested that major Northern Hemisphere glaciation may have been triggered by a shift to an increased seasonal range of sea surface temperatures (SSTs) in the Northern Pacific dated to 2.7 million years ago. In particular, warmer late summer and autumn SSTs in the subarctic Pacific had the effect of increasing the moisture content of westerly airflow, thereby significantly increasing annual snowfall over the northern parts of North America.

Hence, it has become increasingly evident through recent research that a more important global climate threshold was crossed between 3 and 2.5 million years ago than at 1.8 million years ago, and there-fore the majority of workers currently advocate a longer Quaternary chronology and would prefer to place the start of the Quaternary at 2.6 million years ago to correlate with an earlier and more significant geomagnetic transition: the Gauss-Matuyama boundary. At the time of writing, a revision of the Geological Time Scale by the International Commission on Stratigraphy is underway, and it is likely that the base of the Quaternary will be officially changed from 1.8 to 2.6 million years ago.

Reconstructing and Interpreting Past Environments

A remarkable array of evidence is used to infer, or reconstruct, past environments; but similar approaches and assumptions underlie most of the methods. Most of the evidence is stratigraphic in nature, meaning that deposits of sediment (or other material) laid down during the time span of interest are sampled and investigated for clues as to the characteristics of the environment at the time of deposition. If deposition occurred continuously over a fairly long period of time, then the result is a 'stratigraphic sequence' in which changes in the characteristics of the deposit through the sequence are used to infer changes in the environment through time. In addition to the sampling and analysis of deposits, mapping deposits and landforms of similar type and age is another key aspect of palaeoenvironmental study, being necessary for estimating the geographical expression of the past environmental conditions and changes inferred from stratigraphic records.

The types of material studied are diverse, including hard rock, glacial till, fine lake or marine sediments, soil, peat, tree rings, and ice. Samples of stratigraphic sequences are either taken as sections from exposed layers of sediment, for instance in a quarry or along a cliff face, or in the form of cylindrical cores using manual or machine-operated coring devices that penetrate deposits from the surface downwards (or from the outside inwards in the case of tree rings, corals and cave deposits). With most types of deposits, it is possible to make inferences about past environments (that is, to conduct palaeoenvironmental reconstruction) both from the physical characteristics of the deposit itself and from the chemical and biological constituents within the deposit. When used in this way, deposits provide an indirect measure of past environmental or climatic conditions and are often referred to as sources of 'proxy data' because they substitute for the lack of direct measurements during the time period of interest.

The activity of reconstructing past environments is also becoming increasingly linked with efforts to understand the causes of environmental change and to model the environment and climate for predictive purposes. To appreciate the contribution that such studies make towards understanding contemporary environmental change, it is necessary to have some knowledge of the 'systems approach' and to be familiar with the main concepts underlying environmental and climate modelling. Therefore, in addition to discussing the important concepts and assumptions behind palaeo environmental reconstruction, the following sections also provide background to environmental systems and modelling.

The Law of Superposition

When examining a stratigraphic sequence and analysing the proxy data that it contains for palaeoenvironmental reconstruction it must be assumed that material at the bottom of the sequence was deposited first and material towards the top was deposited later. Although obvious, the importance of this assumption for palaeoenvironmental research cannot be overstated. This is known as the 'law of superposition' stating that underlying rock must be older than overlying rock unless the layers of rock have been folded or faulted in such a way that the original sequence becomes inverted. With this simple yet crucial insight the pioneering geologist William Smith was the first to systematically describe relationships between different rock layers (strata), and this made it possible for him to construct the first modern geological map in the world: his geological map of England and Wales.

As with rock, this assumption can be safely applied to other materials such as lake sediments, in which sediment carried into a lake by streams settles on top of older sediment on the lake bottom; or ice on a glacier, in which snow that eventually compacts into glacier ice is continuously added above previously formed ice. However, post-depositional reworking can sometimes disturb or invert a stratigraphic sequence causing problems for interpretation, and this must be borne in mind when choosing deposits to sample. Using the example of lake sediments, the influence of strong currents, lake overturning and high biological activity on a lake bed all contribute to disturbing the sedimentary sequence, and better stratigraphic sequences can be obtained from lakes where these processes are less important. Considering ice, the margins and base of ice-sheets are less suitable for sampling because the ice layers in these places, away from the summit, are more likely to be deformed due to tensional or compressional stresses.

Diagenesis, Equifinality, Provenance, and Taphonomy

Another key assumption for using stratigraphic sequences for palaeoenvironmental reconstruction is that the physical, chemical, and biological properties of the deposit are the product of the past environment *when* the deposit was laid down. However, this may not always be the case due to 'diagenesis' in which post-depositional processes of compaction and chemical alteration change the physical nature of the material from what it was at the time of deposition. This tends to be more of a problem with older deposits that have had more time to be chemically altered, although the degree to which diagenesis interferes with interpretation of the past environment depends very much on the type of material being analysed. In addition to the problems of diagenesis, it is important to be aware of the problem that 'equifinality' can pose for interpreting deposits and landforms. This applies when the distinguishing features of a deposit or landform could arise from more than one set of processes and environmental conditions;

though, like diagenesis, this varies in importance depending on what is being investigated.

When reconstructing past environments it is also often assumed that the deposit under investigation is representative of, and provides information about, the environment *where* the deposit is located. While this is generally the case - for example, organic-rich sediments on a lake bed can reveal much about past conditions and productivity of the lake's ecosystem - this type of assumption is not always as straightforward as it might first appear. For instance, fine, wind-blown dust (loess) can be transported thousands of kilometres from its point of origin before being deposited; hence such deposits owe as much to processes that occurred far away as to local conditions. In fact, it is sometimes very difficult to determine the source area (provenance) of a deposit accurately and to make correct inferences about whether the deposit provides past environmental information at the local, regional, continental, or global scale. There can be uncertainty even when comparing different deposits of very similar nature: for example, the spatial scale represented by evidence preserved in lake sediments may vary with the size and characteristics of the lake, with what part of the lake is sampled, and with atmospheric, hydrological, and ecological conditions prevailing at the time of deposition.

Taphonomy is the study of how a group, or 'assemblage', of fossils found in a particular place came to be there, and environmental reconstruction based on fossils found within a deposit can be complicated by taphonomic problems. Strictly speaking, fossils are chemically altered remains or impressions of organisms left behind in rock, although the principles of taphonomy also apply to the remains of former organisms not yet petrified (for instance plant or animal remains buried in peat or lake sediments), as is often the case in Quaternary deposits. The term 'sub-fossils' is sometimes used to distinguish such remains from remains that are fully petrified.

Taphonomy relates to the problem of provenance described previously in that it is not always easy to determine whether a fossil organism lived where it is found (being *in situ)* or if it was transported from elsewhere before burial. The taphonomy of a fossil assemblage also refers to the conditions which led to preservation, and it is important to be aware that different environmental conditions will influence the extent to which different kinds of organisms are preserved and hence their representation within the assemblage. This means that the proportions of various organisms within an assemblage are unlikely to reflect the actual proportions in the ecosystem when the organisms were alive. For example, shelly organisms such as snails will be better preserved in alkaline deposits, such as marl, than in acidic deposits, such as peat, where calcium carbonate shells are dissolved. On the other hand, cellulose plant material is well preserved in peat because the acidic and anoxic conditions hinder decomposition. By taking into account and correcting for differential transport and preservation, taphonomic studies improve the reconstruction of past plant and animal communities from fossil evidence.

Uniformitarianism and Neo-catastrophism

Another geological concept important for reconstructing past environments is 'uniformitarianism', often phrased as 'the present is the key to the past'. Introduced by Hutton in the late eighteenth century, and further developed in the nineteenth century by Lyell, this is the idea that geological processes observed at present have been occurring in the same fundamental ways for great stretches of time in the past. For example, observations of the deposition of mud in an estuary tell us something about how mudstone is formed and how long it takes to form a layer of mudstone of a certain thickness. This insight completely altered earlier views about the age of the Earth as it became apparent that great spans of time were necessary to explain the formation of the sedimentary rock strata exposed at the surface - not to mention what lies beneath.

While the nineteenth-century concept of uniformitarianism overlooked the sometimes rapid and catastrophic, and at times very different, geological processes that have occurred in the past, it made an essential contribution to our understanding of geological time, as well as our understanding of the development of life. Indeed, Darwin's theory of evolution by natural selection rested on the recognition (made possible by uniformitarianism) of enormous spans of geological time during which natural selection could work. Relatively recent discoveries, from research in the late twentieth century, of rapid changes and major extinction events in the geological record do contrast with the slow and continuous geological changes originally envisioned by Lyell; and new ideas about the environment experiencing long periods of relatively little change interrupted by brief phases of major change are sometimes grouped under the heading of 'neo-catastrophism'.

The term neo-catastrophism distinguishes modern views about the causes of catastrophic change from the earlier (late eighteenth-century) concept of 'catastrophism' in which a much shorter geological timescale was assumed, and major changes were explained by drawing on events described in the Bible such as the Great Flood. There is now much evidence pointing towards rare, but highly devastating, asteroid impacts and volcanic 'super-eruptions' as the most important agents of catastrophic geological change, and the asteroid impact theory for the extinction of the dinosaurs at the end of the Cretaceous Period (65 million years ago), first proposed by Luis and Walter Alvarez in 1980, is perhaps the best-known example of such an event. The fundamental ideas inherent in uniformitarianism and neo-catastrophism are not, however, mutually exclusive. The Earth's geological history should be understood as being shaped by slow processes acting over long periods of time as well as by infrequent, but high magnitude, events.

In the reconstruction of Quaternary environments, uniformitarian principles are often applied when observations of present-day processes are used to surmise how past processes must have worked to create the deposits or landforms under investigation. For instance, such logic is used when modern glaciers are studied in order to infer the behaviour of extinct Pleistocene glaciers in areas that no longer have glaciers today. Observation of the rates of glacial processes acting today also enables inferences to be made about how long it took for various glacial deposits and landforms to form in the past.

This approach is also widely used for interpreting Quaternary environments from the organic remains contained in sediments. Pollen grains, for example, are often abundant and well preserved in terrestrial deposits, and by identifying and counting the different types of pollen and spores in a deposit it is possible to infer past vegetation cover in an area: the Jessen-Godwin pollen zonation scheme for the Holocene and Late Glacial described earlier offers an early example of this approach. By looking at relationships between modern plant communities and climate, it then becomes possible to infer climate conditions at times in the past where evidence for similar plant communities can be derived from the pollen record. This generally works well for the study of the Quaternary Period because most modern plants have existed throughout the duration of the Quaternary, and, in most cases, there has not been enough time for evolution to significantly change their climatic tolerances. Of course, using modern plant communities as analogues for past vegetation becomes less valid the further removed in time the past vegetation is from the present. Another potential problem with this type of analogue approach is that even in very recent geological time (just thousands of years ago) there are cases when climatic and environmental conditions were so different from at present that specially adapted plant communities existed for which there are no equivalent modern analogues.

In addition to uniformitarianism, insights of neo-catastrophism also contribute to understanding and reconstructing past environments. Super-eruptions such as those of Yellowstone at 2.1, 1.3 and 0.65 million years ago and of Toba on Sumatra 73 500 years ago caused major environmental and climatic changes on a scale well beyond what has been observed from volcanic eruptions in historic times. In addition, the operation of 'non-linear processes', by which minor climate perturbations trigger disproportionately large environmental changes, is being increasingly recognized. When this happens, a minor change to the system that manages to push it over the threshold is amplified by positive feedback mechanisms until conditions settle into a new state of stability in which the climate and environment are significantly different from what they were before.

Environmental Systems

An important paradigm for understanding the Earth's climate and environment is the idea of environmental systems, and within this conceptual framework, uniformitarian and catastrophic-type changes can be seen as different modes of operation within the system. Seeing the Earth as a set of interconnecting 'environmental systems' derives from general systems theory which explains phenomena in terms of flows and transformations of energy and matter. There is a seemingly infinite variety of natural systems ranging widely in type and scale: from the biological system of a single cell to the physical system governing circulation of the entire atmosphere. Yet the operation of all systems can be understood according to a set of fundamental principles. A throughput of energy is of course essential for driving the system, and the state (or characteristics) of the system at any one time is a result of interactions between its various component parts. Over time the system may change in a variety of ways depending on changes in the balance between inputs and outputs of energy and matter, as well as on changes in the internal functioning of the parts that make up the system.

In systems terminology, a closed system is one in which all the matter is circulated inside the system with only inputs and outputs of energy; whereas an open system has inputs and outputs of both matter and energy. The environmental systems paradigm sees the Earth's total environment as being controlled by the interaction of numerous open systems, such as atmospheric, hydrological, geological, and biological systems, and changes in the Earth's environment cannot be explained without an understanding of how these component systems interact with each other. One example of the way such open systems interact is in the transfer of H_2O. H_2O existing as liquid on the Earth's surface is transferred to the atmosphere because the input of solar energy causes evaporation. Later, when vapour condenses in the atmosphere to form cloud droplets, the energy originally used to evaporate the water is released as latent heat (warming surrounding air) due to the phase change from vapour back to liquid. Provided that droplets grow large enough, the H_2O will then leave the atmospheric system as precipitation and, if over land, enter a drainage basin system. From there it could follow any number of pathways, perhaps being taken up by vegetation or returning to the ocean via run-off where it may eventually be evaporated again to re-enter the atmosphere. The hydrological cycle is just one of many different ways by which environmental systems interact in terms of material and energy transfer. What makes understanding environmental change so complex is that the component systems themselves are governed by the interactions of smaller-scale sub-systems, which in turn, contain their own sub-systems and so on.

Systems analysis was brought into the earth sciences in the mid-twentieth century, enabling a more integrated view of the environment and a better understanding of how it changes. An important offshoot of this was the recognition of feedback mechanisms in environmental change. The idea of positive feedback has already been introduced in the context of threshold changes in which an initial change in the system leads to further changes that reinforce the initial change - causing a 'snowballing' effect. On the other hand, negative feedback mechanisms are processes that dampen the initial change, thereby working to bring the system back to its original state. Negative feedbacks are equivalent to homeostasis: the processes by which an organism regulates internal conditions (such as body temperature) in response to external stimuli. The environmental systems paradigm recognizes that both positive and negative feedback mechanisms operate when systems interact with each other, and that fluctuations in environmental conditions can be viewed as the result of the interplay between different feedbacks.

Because of the complexity of interactions within environmental systems, environmental and climatic conditions can respond to changes in a particular variable (or forcing mechanism) in a number of different ways. Environmental responses can be proportional, insensitive or sensitive to the forcing mechanism, as well as abrupt in the case of the threshold-type response. In proportional change, environmental conditions alter to an extent in line with the change in the forcing mechanism. The insensitive condition describes a situation where a variable changes, but the system is largely immune to the applied forcing. In the sensitive condition environmental conditions respond strongly to the applied forcing, and in a threshold response there is a rapid shift in the system from one state to another.

Within the context of Quaternary studies, the application of systems theory and an understanding of feedback mechanisms and different system responses are essential for going beyond description to explanation of palaeoenvironmental changes inferred from the evidence. Discovering relationships between past changes in a forcing variable and past changes in the environment is of great importance for understanding how the environment may respond to similar changes in the future, and it is through this type of systems analysis that palaeoenvironmental studies are most relevant to contemporary debates about human-caused environmental changes such as global warming.

Environmental Modelling

Predictions of environmental change require the construction of models, and reconstructions of past environments are playing an increasingly important part in the design and testing of such models.

In this sense, the term 'model' refers to a simplified representation of a natural system that is used to investigate the changes to some aspect of the environment that result when certain variables (environmental factors) are changed. When designing such a model, modellers seek to represent only the interactions within the natural system that are thought relevant to the variable of interest while ignoring those processes that are believed to be negligible. Because of the enormous complexity of the environment, and our imperfect knowledge of all of the processes operating within it, it is impossible to design a model that represents all interactions perfectly; and this is why all models, even the most complicated, are simplifications of reality resting on certain assumptions about the workings of the system. Assumptions about what processes should be included or excluded, and about the way in which the system will respond to forcing, lay models open to criticism; although as understanding of natural systems improves through experiment and observation, so too does the accuracy of the assumptions used in model design.

Models can be qualitative or quantitative, with the former useful for visualizing connections between factors and for drawing general conclusions about cause and effect, and the latter necessary for precise predictions of the magnitude of change. General Circulation Models (GCMs) used for predicting the amount of atmospheric warming that will occur in response to human-caused emissions of greenhouse gases are the best-known and most complicated of the quantitative type. The outputs of different GCMs from various universities and institutions around the world form the basis for the Intergovernmental Panel on Climate Change (IPCC) predictions of climate change over the next century. These state-of-the-art mathematical models are run on the world's fastest super-computers, yet a single GCM simulation of atmospheric behaviour over a timescale of years can take many days to compute.

GCMs represent the atmosphere as a number of interacting grid boxes across the surface of the globe, and the current generation of models can represent the whole atmosphere with boxes measuring 250 by 250 km across and 1 km high. To start the model, each box has to be assigned initial characteristics (for example, temperature, humidity, cloudiness, windspeed, etc.) and when the model is run, a factor in the boxes, such as carbon dioxide concentration, is changed and equations are used to calculate the effect on other variables as well as how each box will influence adjacent boxes. Typically the model will recalculate every variable in every grid box on a simulated timescale of once every half-hour. It is hard to imagine the enormous number of calculations required to compute such a model for even a one-year simulation at this spatial and temporal resolution. GCMs have become increasingly sophisticated in recent years as the atmospheric models have been 'coupled' with oceanic models (the Atmosphere-Ocean General Circulation Models or AOGCMs), and some GCMs also factor in processes involving variation in the land surface, the biosphere and the cryosphere (the world's ice).

Palaeoenvironmental reconstruction contributes to modelling efforts in two main ways. First, proxy data from stratigraphic evidence sometimes reveal environmental changes in the past of a type or magnitude previously unimaginable within the prevailing orthodoxy about how climate and environmental change works. This throws up new challenges to theoreticians because assumptions about how environmental systems interact must be revised in light of the new evidence. The challenge then passes to modellers who must find ways to incorporate previously unknown processes into their models so that they more closely represent the real world. An excellent example of this type of contribution concerns the discovery of abrupt and high-magnitude climatic changes during the transition from the Late Glacial to the Holocene. It became apparent in the 1990s from various forms of proxy data derived from terrestrial, marine, and ice-sheet contexts that Atlantic Ocean circulation can fluctuate dramatically over short timescales in response to freshwater input causing major climatic changes of a speed and scale that might have been thought impossible if it were not for the palaeoenvironmental evidence. In response, climate modellers have been redesigning GCMs to take into account this new knowledge of how ocean, ice, and atmosphere can interact, and these developments are the root of widely publicized concerns about global warming actually causing north-west Europe to cool if increased meltwater from polar snow and ice disrupts the North Atlantic's warm currents.

The other way in which palaeoenvironmental reconstructions contribute to modelling is by providing snapshots of the climate and environment at certain times in the past that model simulations can try to replicate. This is important in the process of 'validating' GCMs. If a climate model can adequately reproduce climatic conditions at a specific time in the past when the 'boundary conditions' of the model are altered to reflect the past background characteristics (such as the distribution of land, sea, ice-sheets, and concentrations of greenhouse gases and dust in the atmosphere) then more confidence can be placed in the model's ability to predict change in the future. This is an important reason for some of the large-scale, coordinated efforts in palaeoenvironmental reconstruction undertaken in recent decades. Two such collaborations that pioneered this approach, and involved researchers from many institutions, were the CLIMAP project and COHMAP (Cooperative Holocene Mapping Project). The CLIMAP project pooled temperature proxy data from marine sediment cores to make maps of sea surface temperatures (SSTs) across the world as they were at the height of the last glacial phase approximately 18 000 years ago. This provided an essential boundary condition (the SSTs) for incorporation into GCMs that try to simulate the glacial climate. The COHMAP project was concerned with modelling the changes in global climate that occurred from 18 000 years ago up to the present day in 3000-year 'time slices'. With this type of exercise, the ability of the model to simulate the climate at any one time can be tested against the climate reconstructions based on Quaternary proxy data.

In summary, collecting proxy data and constructing models are both important activities for understanding environmental change, and each has much to contribute to the other. While proxy data may reveal processes previously unaccounted for in models, modelling studies may in turn draw attention to processes and factors initially overlooked in the palaeoenvironmental record, thereby stimulating further development and new directions in palaeoenvironmental research.

References

Edwards, Paul Geoffrey; Miller, Clark A. (2001). *Changing the atmosphere: expert knowledge and environmental governance*. Cambridge, Mass: MIT Press.

Emanuel K (August 2005). "Increasing destructiveness of tropical cyclones over the past 30 years" (PDF). *Nature* 436 (7051): 686–8.

McKibben, Bill (2011). The Global Warming Reader. New York, N.Y.: OR Books.

Ruddiman, W. F. (2003). "The anthropogenic greenhouse era began thousands of years ago". *Climate Change* 61 (3): 261–293.

Ruddiman, W. F., Vavrus, S. J. and Kutzbach, J. E. (2005). "A test of the overdue-glaciation hypothesis". *Quaternary Science Review* 24 (11).

Schmidt, G. A., Shindel, D. T. and Harder, S. (2004). "A note of the relationship between ice core methane concentrations and insolation". *Geophys. Res. Lett.*31 (23).

Wagner, Frederic H., (ed.) *Climate Change in Western North America: Evidence and Environmental Effects* (2009).

2

Stresses on Ecosystems

A recent research published in the journal *Nature* defines,, for eight earth biophysical subsystems or processes, "safe" threshold boundaries that should not be exceeded to avoid major environmental disruptions. Already several thresholds appear to have been exceeded, including: climate change (as measured by excess CO_2 concentration in the atmosphere); the rate of biodiversity loss (terrestrial and marine); and interference with the nitrogen cycle (N_2 is removed from the atmosphere and converted to reactive nitrogen for human use mainly for agriculture). Interference with the global phosphorous cycle, ocean acidification, global freshwater use, and change in land use are approaching their thresholds. No boundaries have yet been established for chemical pollution and atmospheric aerosol loading. Stratospheric ozone depletion is a noteworthy exception, where international environmental cooperation has yielded progress in reversing a negative trend.

Climate Change: CO_2 emissions have been rising steadily: Since 1751 approximately 329 billion tons of carbon have been released to the atmosphere from the consumption of fossil fuels and cement production. Half of these emissions have occurred since the mid 1970s.

Biodiversity Loss: The Millennium Ecosystem Assessment evaluated the ability of ecosystems to provide ecosystem services such as the provision of goods (e.g. food, water, fibre and fuel), the regulation of natural systems (e.g. climate, water and disease), cultural benefits (e.g. aesthetics, spiritual, recreation and education), and other supporting benefits (e.g. primary production and soil formation). The assessment concludes that humans have changed ecosystems in unprecedented ways over the last 50 years. Sixty percent of the world ecosystem services have been degraded, which is putting such strain on the natural functions of Earth that the ability of the planet's ecosystems to sustain future generations can no longer be taken for granted.

The International Assessment on Agricultural Knowledge, Science and Technology for Development (IAASTD) finds that inappropriate fertilizer application has led to eutrophication and large dead zones in a number of coastal areas and some lakes, and inappropriate use of pesticides has led to groundwater pollution and loss of biodiversity. In addition, 1.9 billion hectares (and 2.6 billion people) are affected by significant levels of land degradation. The area of drylands has been growing steeply since the 1980s as a share of total land area. Very degraded soils are found especially in semi-arid areas, areas with high population pressure and in regions that are undergoing deforestation. Biodiversity is highly correlated to the number and health of ecosystems. Vertebrates have been the most affected, with 10% of species threatened, followed by plant species with 3%.

Interference with the nitrogen cycle: Nitrogen is needed to grow food but because of the inefficiencies of nitrogen uptake by plants and animals, only about 10 to 15 percent of reactive nitrogen is taken up. The rest is lost to the environment and injected into the atmosphere by combustion. This nitrogen pollutes water sources and the world's oceans, harming marine ecosystems, and contributes to global warming. Agricultural runoff and the burning of fossil fuels have boosted the supply of reactive nitrogen in the open oceans 50 percent above the normal range.

Roughly four times more nitrogen fertilizer was applied in 2000 than in 1960, and applications have increased steadily since then. This has led to increased deposition and change in the N cycle in various ecosystems. Evidence suggests very high application rates contribute to soil degradation.

Ecological Footprint analysis — which compares humanity's ecological impacts to the amount of biologically productive land and sea available to supply key ecosystem services (food supply, fibre, habitat, carbon storage, etc.) — finds that the global economy started exceeding the planet's biocapacity in the 1980s, and overconsumption of resources has increased since then. The single largest demand humanity puts on the biosphere is its carbon footprint, which has increased tenfold since 1961. The Ecological Footprint exceeds the earth capacity to regenerate by 30%.

Alternatively, 1.3 planets would be needed to stay within the planet's carrying capacity.5 This is another, simplified way of picturing planetary boundaries and ecosystem thresholds.

Under a business-as-usual scenario, 2 planets would be required by 2030 to support the world's population. This assumes a continued unequal world with 15% of the population using 50% of the resources. World Wildlife Fund (WWF) estimates that three planets would be needed now if every citizen adopted the UK lifestyle, and five planets if they adopted the average North American lifestyle. Regional differences are driven by differences in geography and climate, lifestyle, affluence, population and technological sophistication. At the regional and national level, a comparison of ecological footprints with biocapacity indicates whether countries and regions live within their biological carrying capacity (or "import" such capacity from the rest of the world).

The net balance of a country's footprint depends on its consumption compared to its biocapacity per capita.5 Russia, Canada, the United States, Brazil, Australia, India, and Argentina have the largest biocapacity. Three of these have ecological footprints that exceed their biocapacity because of large population (China and India) or high consumption per capita (the US). Countries exceeding their biocapacity went from none in 1960 to 24 countries at present. North America, with the largest biocapacity per capita, still exceeds its biocapacity, while Africa with a relatively small biocapacity per capita and 902 million people has a biocapacity reserve.

Water Foot Print: Local and regional imbalances between water availability and growing demand are a growing concern globally. About 50 countries are already experiencing moderate to severe water stress all year round, while many others have water stresses during part of the year. Many lakes and rivers have dried out following extensive extraction and damming to irrigate agriculture. Water-intensive agricultural products include meat, dairy products, sugar and cotton. Water use efficiency is often low, partly as a result of low water pricing or even subsidies, irrespective of scarcity. Lack of awareness of simple water saving measures among farmers and the use of water inefficient technologies also contribute. The water footprint of a country, an indicator introduced in 2002, measures the volume of freshwater used to produce the goods and services consumed by the inhabitants of a country, calculated over the product's entire global supply chain. It is composed of the internal and external water footprint.

Internal footprint refers to the appropriation of domestic water resources and external footprint to the appropriation of water resources in other countries. Worldwide, 27 countries have an external water footprint which accounts for more than 50 per cent of their total water consumption, that is, they depend heavily on water-embodied trade (or 'virtual water' trade). Major rivers and lakes, which serve as sources of drinking water, are increasingly polluted from both industrial and agricultural sources, including nitrogen. Generally, river pollution increases with nitrogen use intensity, though with considerable dispersion.

Ecosystem Responses to Global Climate Change

The responses of functional groups of plants to altered climate can be predicted from their known responses to fluctuations in soil resources (nutrients and water) and the expected effect of climatic change on these soil resources.In this section we can see, how environmental tolerance of plants can be used to predict their responses and those of ecosystems to climatic change. There are two basic sources of information on which to base these predictions: (i) the relationship of plant distribution to present environment and (ii) observations on the responses of plants to natural or manipulated changes in environment.

Patterns of Vegetation Distribution

Climatic extremes appear more important than climatic averages in predicting patterns of vegetation distribution. The poleward limits of many tree species are determined by frost sensitivity and freezing tolerance. For example, –40°C is the lower limit of supercooling for most ring-porous hardwood trees. Below this temperature intracellular freezing occurs, killing the cells and therefore the organism. The northern limit of these tree species correlates closely with the record low temperature of –40°C. Conifers, which have a different mechanism of freezing tolerance, can withstand the temperature of liquid nitrogen and have a correspondingly more extreme latitudinal limit to their distribution. Similarly, chilling-sensitive broad-leaved tropical evergreen trees are killed at 15°C, and their poleward limit coincides with record low temperatures of 15°C.

Woodward recognised several major physiognomic types of vegetation: conifer forest, deciduous forests, broadleaved evergreen forests, shrubland, and herb-dominated vegetation, based on the assumptions that climate governs the low-temperature or low-moisture limit of distribution, and that the mesic limits of distribution are determined by the environmental tolerance of a competitively superior but less tolerant physiognomic type, Woodward successfully predicted the global distribution of most major physiognomic types of vegetation.

These observations suggest that the lowtemperature or low-moisture limit of a physiognomic type is governed by the fundamental niche (i.e., physiological tolerance) of the plants, whereas additional limits of plant distribution may be determined by other, more complex factors, probably mediated by competition. These patterns also indicate that extreme events, which are a function of environmental fluctuation, are more useful than average conditions in predicting northern limits of physiognomic types. As climate changes, the location and frequency of these extreme events will certainly change, leading to changes in distribution of biomes. Thus, understanding of patterns of environmental variability are critical to predictions of future vegetation distribution.

Although extreme events predict distributions of general physiognomic types, they are inadequate to predict finer distribution patterns of species or functional groups of species, i.e., groups of species which show similar responses to change in environment. The greater diversity of response by species than by physiognomic types occurs because each species has a unique range of environmental conditions under which it occurs, which seldom coincides precisely with their limits of physiological tolerance.

Correlation of the current distribution of plants with their current average climatic conditions has been the main basis of explaining past patterns of vegetation change and predictions of future vegetation distribution. These studies have led to several important generalisations:

(i) Each species shows a unique pattern of distribution with climate and responds most strongly to different patterns of climatic factors, so that complex climatic changes cause species to migrate with different patterns and to form new associations. Thus, current plant communities are temporary associations among species that last only hundreds or a few thousand years. These communities did not exist as entities in the past, and there is no reason to expect their continued coexistence in the future. Because of the highly individualistic response of species to climate and because this response depends on interactions with other species in the community, it is difficult to determine which aspects of climate are particularly critical to the distribution of a species, making predictions of future distribution difficult.

(ii) Changes in the distribution of a species may lag significantly behind climatic changes where dispersal limits the rate of species migration. Thus, predictions of the future response to climate requires some understanding of factors governing the regeneration phase.

(iii) Projections of the future response of vegetation to climate are quite sensitive to availability of soil resources, so that knowledge of climate alone is inadequate to predict future species distribution.

(iv) Finally, in complex and more diverse communities such as dry or wet tropical forest there are so many important species that we can never hope to understand their climatic controls well enough to predict future vegetation changes. What is needed is a simplified classification of species into functional groups whose environmental controls can be predicted from general ecological principles.

Vegetation with Soil Resources

Given the difficulties of predicting direct species reponses to climate and the importance of soil resources in mediating plant responses to climate, the correlation of vegetation with soil resources may provide an effective means of predicting the response of functional groups of plants to climate change. Soil resource availability will respond in predictable ways to altered climate. In cold climates, increased temperature will stimulate microbial activity and nutrient cycling, thus increasing the availability of commonly limiting nutrients such as nitrogen and phosphorus. In temperate mesic-to-dry climates, increased temperature and potential evapotranspiration combined with moderate changes in precipitation will decrease soil moisture in mid-continental regions.

In tropical coastal dry climates, moderate increases in temperature with high variations in precipitation are expected, as observed in the past. Rates of nutrient cycling correlate closely with availability of most soil resources. Forest clearing for agriculture creates early successional habitat with associated increases in light availability. Local and regional variations in climatic projections and soil fertility provide logical bases for refining these predictions.

The major point is that patterns of change in soil resources can be predicted from defined scenarios of climatic change. Soil moisture responds readily and predictably to changes in climate. However, considerable research is necessary to determine to the magnitude and timing of the response of soil fertility to changes in climate. At present it is also difficult to predict how changes in the seasonality of climate affect soil resources and the plant phenological patterns necessary to exploit these soil resources effectively.

Changes in availability of soil resources lead to predictable changes in the types of plants that can be expected. In brief, high-resource environments support a high relative growth rate (RGR) through high capacities for photosynthesis and nutrient uptake, which in turn require high tissue-nitrogen concentrations. Continued high rates of resource capture require high rates of root and leaf turnover, a process that can be supported at relatively low cost in a high-resource environment. Conversely, in low-resource environments there are inadequate resources to support rapid growth, so plants are constrained to grow slowly and are likely to achieve a small size. Because of low tissue-nutrient concentrations, plants in low-resource environments have low potentials to photosynthesise, transpire, and absorb nutrients. Plants in these environments have high root-to-shoot ratios to maximise capture of scarce soil resources. Tissue-turnover rates are low, causing tissue nutrients to be retained for a long time but also requiring effective chemical defense against herbivores and pathogens. These chemical defenses reduce litter quality and reinforce the low nutrient availability in these sites.

Experimental evidence supporting these predictions has emerged mainly from studies in temperate ecosystems, although recent studies in tropical rain forests suggest similar patterns with respect to relative growth rate, root-shoot ratios and mineral nutrient requirements. Less is known about the highly diverse tropical deciduous forest in which the availability of soil resources is restricted by seasonal drought. In these forests tree-seedlings occupying resource-rich disturbed areas (e.g., *Heliocarpus pallidus)* show high relative growth rates, high demands for mineral nutrients and a low root/shoot ratio. These species also tend to be more sensitive to water and mineral nutrient stress.

Environmental Fluctuations

Differences in resource requirements determine plant response to environmental fluctuations. Plants adapted to high availability of soil resources are sensitive to changes in nutrient supply and other resources such as light and water. In these habitats plants respond plastically to localised depletion zones around root systems through morphological changes in roots, resulting in reallocation of absorptive surfaces from depleted zones into the resource-rich areas. This system of patch exploitation requires shoots and roots with a short life span and high rate of tissue turnover.

A high degree of morphological plasticity associated with active foraging will be of selective advantage only where it promotes access to large reserves of light, water, and mineral nutrients. Conversely, in habitats where productivity is chronically resource-limited, there is less morphological plasticity. On infertile soils we expect that resource capture and survival will depend on successful exploitation of resource pulses. In unproductive habitats, in which growth is frequently uncoupled from resource capture, plasticity would involve reversible physiological changes rather than reallocation of biomass to facilitate exploitation of resource pulses. These generalisations come from screening programmes which document the responses of plants to fluctuations in resource supply.

Environmental responses of communities

In artic tundra, when light temperature, and nutrients were manipulated to simulate patterns expected with climatic change, each species initially showed a species-specific, unique response to our manipulations, which was difficult to predict from general principles. However, after nine years of treatment, there continued to be species-specific, unpredictable responses to temperature, but relatively predictabl responses to nutrients: For example *Betula nana* and other rapidly growing deciduous shrubs increased growth in response to nutrient addition, whereas *Ledum palustre* and other slowly growing evergreen species responded negatively to nutrient addition.

Similarly, all species except the most shade-tolerant understory species responded negatively to reduction in light intensity. These results suggest several important conclusions. First, the resource responses observed in the field were consistent with predictions of laboratory experiments in that rapidly growing species responded positively to nutrient addition but slowly growing species responded negatively to this improvement in resource supply, presumably through changes in competitive balance. Secondly, it was easier to predict responses to altered resource supply than to altered climate.

The relatively minor increase in nutrient availability that occurred nine years after initiation of our temperature treatment suggests that our experiment was too short for climate to strongly alter soil resource supply. A third result of these experiments was that manipulations which benefited some species reduced the biomass of other species. Therefore, the overall production of the ecosystem was affected much less than the productivity of individual species or functional groups. Thus, the contrasting responses of individual species and functional groups buffer ecosystem processes such as production and nutrient cycling, rendering them relatively resistant to environmental change.

The buffered response of ecosystem processes is also seen in an unmanipulated tundra ecosystem sampled over a period of years. Over a series of years in which the productivity of individual species varied 2-8-fold, there was no significant variation in productivity of the total community. Addition of water and nutrients to grassland also gives predictable responses by functional groups of species. Rapidly growing forbs and grasses responded positively to nutrient and water addition, whereas the more slowly growing cactus responded negatively. Moreover, when grassland production was examined over a series of years, the productivity of individual species varied much less than did the productivity of the community as a whole, again indicating the extent to which ecosystem processes are buffered by the compensatory responses of individual species.

Climate Change and Ecosystem Functioning

A functioning ecosystem is made up of animals and plants interacting in a myriad of different ways. Animals are also important members of ecosystems, playing a major role in maintaining ecosystem function through the ecosystem services they provide (e.g., pollinators, seed dispersers, insect predators, decomposers, etc.). Animals can be intensely affected by weather and climate, perhaps to an even greater extent than plants. Consequently, any examination of the impacts of abrupt climate change on ecosystem functioning needs to look at the animals that provide critical ecosystem services.

Animals can also be effective barometers of their environment. Their overall mobility and the physiological constraints imposed on many of them by temperature and moisture make them effective indicators of climatic changes. Indeed, many different taxa in various parts of the world are already exhibiting recognisable changes, such as poleward and elevational range shifts, and changes in the timing of events like breeding. Many factors (e.g., habitat conversion, and climate change) put pressure on animals. Nevertheless, the balance of evidence suggests that, for those animals exhibiting significant large-scale patterns of changes, the most consistent explanation is recent climatic change. Like the proverbial "canaries in the coal mine," wildlife appears to be providing an important early indicator of how ecosystems might respond to the discernible human impact on climate that is contributing to its change.

Climate and climate change are strong drivers of biotic systems; the distribution, and survival of many species being moderated by climate. Although species have responded to climatic changes throughout evolutionary time, the primary concern is the projected rapid rate of change. High species richness appears to be related to stable conditions, while abrupt impoverishment of species has occurred during times of rapid change. When changes in climate are gradual, species may have enough genetic variability to allow them to adapt. However, the current projected rate of temperature change is likely too fast for many species to naturally adapt successfully to climatic change. If the climate changes abruptly then the likelihood of natural adaptation is quite small.

The distributions of many plants and animals are limited by temperature, precipitation and soil moisture. Overall, the geographic range of plants and animals will tend to shift poleward and/or upwards in elevation in response to temperature changes with some modification by precipitation changes. Range shifts of wildlife populations will be dependent upon factors such as the availability of migration corridors, suitable habitats and the concurrent movement of forage and prey.

Range shifts in plants will be dependent upon factors such as soil types, migratory pathways, seed dispersal mechanisms and pollinator availability. Further complicating potential range shifts will be other landscape changes such as roads, cities, and habitat fragmentation, all of which can present significant barriers to species range shifts. As temperature and rainfall patterns change, some species may benefit and some may decline. However, precise predictions of ecological change are not possible due to the scale and accuracy of current climate models. Predicting effects on plants and animals is further confounded by a lack of information concerning species-level response, interactions among biotic and abiotic components of ecosystems, and uncertainties related to non-climate stresses on ecosystems. The best available evidence from paleoclimatic studies, and current models and observations suggests that the individual physiological responses of plant and animal species to temperature and moisture make it unlikely that species will respond in the same manner to climate change. It is unlikely that plant and animal species will respond in the same manner to climate change with each species potentially moving independently. Thus, ecological communities as we now know them will look different in the future. Indeed, there is sufficient evidence to indicate that many animals and plants have already begun to change in response to observed climatic changes. These changes could ultimately lead to a decoupling of species and a disruption of ecosystem services; the faster the rate of change, the greater the likelihood of ecosystem disruption.

Major Changes in Species Ranges

Early estimates of animal range changes simply assumed that the ranges would change in concert with gross changes in habitat types. While often not the preferred technique, this level of analysis does provide a useful first approximation as to how the ranges of many species may potentially change. Ideally, analyses need to go much farther with a more useful first approximation being an examination the potential changes of species ranges individually (as opposed to groups) in response to abiotic factors. Better still are the few analyses that look at both abiotic factors and habitat.

Current models project an increase in sea-level of 3.5 – 35 inches (.09 - .88 m) by 2100. Some coastal areas are also subsiding, contributing to a loss of coastal wetlands and tidal mudflats important for shorebirds. This loss of habitat and reduction in foraging area could potentially lead to a redistribution of shorebird concentration sites or to declines in shorebird numbers. While Bolivar Flats, a globally important shorebird stopover site, is expected to potentially gain tidal mudflats (habitat migration) by 2100 it could lose between 38% and 81% of its tidal mudflats between now and 2050.

The overall amount of change in tidal mudflats is dependent on many things, such as local subsidence, the amount of temperature change/sea-level rise, and the ability of habitats to migrate inland in the face of barriers to migration such as roads and development. Human intervention to sea-level rise, such as the construction of barriers, may actually exacerbate the loss of shorebird habitat at Bolivar Flats. It is not only tidal mudflats that are at risk from sea-level rise. The capacity of this system to respond to sea-level rise is highly dependent upon the rate of change. There needs to be few to no barriers to habitats migrating inland (e.g., coastal development) and the slope of the shoreline needs to be situated such that these new habitat types can form as the sea level rises.

Models project that over the next century there could be major changes in the location of many vegetative communities in the United States and elsewhere. For example, alpine tundra could largely be lost from the conterminous U.S., likely leading to population declines in White-tailed Ptarmigans. In the West, much of the Sonoran Desert could ultimately become arid grassland. In the Midwest, savanna woodlands and shrub woodlands could replace some grasslands. In the Southeast, southern hardwoods could replace many areas that are currently southern pine forest.

In the Northeast, a northern mixed forest type could replace conifer forests and the maple-beech-birch communities could be replaced with oak-hickory. In Alaska, temperate evergreen and boreal conifer forests could ultimately replace large areas currently classified as taiga/tundra. These habitat changes will ultimately lead to a reorganisation of species composition and ecosystem functioning. The ability of habitat types to "convert" to a different habitat type will depend on both the magnitude and the rate of change as well as on the factors previously mentioned.

As many tree species are long-lived and migrate slowly they would be expected to slowly colonise to the north of their range (or up in elevation) while at the southern edge of their range their rate of reproduction slows and finally stops. Even once a species stopped reproducing the habitat may not undergo much compositional change until the existing community dies out. Thus, it could take decades to centuries for species in some vegetative communities to be replaced by others. As increased temperatures and drought stress plants they become more susceptible to fires and insect outbreaks. These disturbances will likely play a large role in the conversion of habitats from one type to another. There could very well be instances where the existing plant communities are lost to disturbance but climatic conditions and migration rates limits the speed by which a new vegetative community replaces the original. Thus, some areas may transitionally be replaced by grasslands, shrublands and, especially, by invasive species.

The probability of these transitional habitats may very well be increased with abrupt climate change. The faster the change in climatic conditions the less time there will be for existing species to be replaced by more physiologically suitable species. Additionally, a more rapid rate of climatic change may put additional stress on existing species by reducing the potential for any phenotypic adaptation to the changed conditions.

Changes in Bird Distributions

Summer bird ranges have often been assumed to be tightly linked to particular habitats. This is often only partially true. While certain species are usually only found in specific habitats, others are more flexible in their habitat use. Species found in a particular habitat type throughout their summer range may not be found in apparently suitable habitat north or south of their current distribution. Birds are also limited in their distributions by their physiology and food availability.

The link between physiology and the winter distributions of many species is well established and recent research shows that physiology plays a role in limiting summer distributions as well. Often, the choice of a specific habitat may be to provide a microclimate suitable for a species' physiology. While habitat selection, food availability, and competition may all play a role in influencing local distributions of a given bird species, looking at a species' overall distribution often suggests that a climatic component may play a significant role.

To determine how summer distributions of North American birds might change, it is first necessary to ascertain whether there is any association between distributions and climate. If an association exists, then an examination of projected future climates can be used to see how the climatic ranges of birds might change. Building on earlier work that found that many winter bird distributions were associated with climate, models of the association between climate and the distributions of more than 200 species of North American birds have been completed. These bird distribution models have also been used to look at potential range shifts for a doubling of carbon dioxide. A more complete explanation of methods used to develop the models and maps can be found in Price. Similar results have been obtained by other authors using similar techniques to look at other taxa around the world.

What these studies show are how areas projected to have the proper climate for a species, or climatic range, may shift as the climate changes. They do not deal with the rate of climatic change. However, given that many species of birds regularly shift their range in response to climate variability it seems likely that many of these changes might occur under abrupt as well as with more gradual climate change scenarios. While the results of these models often cannot be used to look at the fine points of how a given species' distribution might change, they can provide an impression of the possible direction and potential magnitude of the change in the suitable climate for the species.

Ultimately, the greatest impact on wildlife and vegetation may not come from climate change itself, but rather from the rate of change. Given enough time, many species would likely be able to adapt to climatic shifts, as they have done in the past. However, the current projected rate of warming is thought to be greater than many species will be able to genetically adapt to. Under an abrupt climate change scenario the probability of adaptation, even behavioral adaptation may be reduced. How quickly these distributional changes might occur is unknown. The rate of change will largely depend on whether limits to a given species' distribution are more closely linked with climate (especially temperature), vegetation, or some other factor. The rate of change will also likely be tied to the rate of change of the climate itself.

If the climate changes relatively slowly, then species may be able to adapt to the new climate. However, many changes could (and are) occurring relatively quickly. One study found that the average latitude of occurrence of some species of Neotropical migrants (e.g., Cape May, Bay-breasted, Golden-winged and Bluewinged Warbler) has already shifted significantly farther north in the last 20 years, by an average distance of almost 60 miles (100 km). The center of the range of Prothonotary Warbler is farther north, and more BBS routes are now detecting this species in the northern part of its range than the southern portion. The arrival date of many species of migratory birds on the Upper Peninsula of Michigan was found to be 21 days earlier in 1994 than in 1965. This includes long-distance migrants like Rose-breasted Grosbeak, Black-throated Blue Warbler, Nashville Warbler and Barn Swallow and short-distance migrants like Eastern Bluebird and Red-winged Blackbird. Many other species have been found to be arriving and breeding earlier, not only in the US but in Europe and elsewhere. A faster rate of climate change would potentially lead to even greater impacts on biodiversity.

Shifts in individual species' distributions are only part of the story. It is also unlikely that ranges of coexisting species will shift in concert. Bird communities, as we currently know them, may look quite different in the future. For example, one study of grassland species projects that future communities may only contain 50% of the species currently found there. Climate change may shift the range of Neotropical migratory birds and change the avifauna in the conterminous United States and southern Canada.

For example, under climate change conditions projected by Canadian Climate Center, the Great Lakes region of the United States could see a potential gross loss of 53% of the Neotropical migrants currently found in the region's states. These losses would be somewhat offset by birds colonising from outside the region so the net change would still be 29% fewer species than are currently found there. Other analyses have posited an increased risk of extinction for some threatened and endangered species, including the Golden-cheeked Warbler. Similar work has looked at how the ranges of endemic species may change in response to a gradual climate change scenario finding that as many as xx% of species may potentially face extinction.

Potential Rates of Change of Birds and Plants

As species move, they will face new prey, predators and competitors. So-called "optimal" habitats may no longer exist, at least in the short term. The potential rates-of-change of birds and the plants that shape their habitats are often quite different. While many birds may be able to respond quickly to a changing climate, some plant ranges may take from decades to centuries to move. Thus, some species may face a bottleneck with physiological limits pressing them to move but having no habitat available to move into. Only recently has work been done looking at how climatically induced changes in the ranges to organisms might, in turn, affect the ecosystems of which they are a part.

An ecosystem might continue to function if species are removed. However, how well it functions depends on the role of the lost species. As more species are removed, the probability increases that the system will break down. So, one important concern of researchers looking at the potential impacts of climate change on ecosystems is whether or not ecosystems will become decoupled. That is, will the species within an ecosystem respond differentially to a changing climate? The ecosystem services birds (and other species) perform contribute significantly to a healthy ecosystem. These services include seed dispersal, pollination, decomposition of organic matter, primary productivity, and predation on insect pests. The loss of ecosystem functioning is likely a much greater source of uncertainty (and 'surprises') than the changes in ranges of the species themselves.

Predator-prey interactions

The eastern spruce budworm (Choristoneura fumiferana) is a common, native planteating insect of boreal forests in eastern North America. Its preferred foods include balsam fir (Abies balsamea) and several spruce species (*Picea spp*.). Budworms are estimated to annually defoliate millions of hectares in the United States and Canada. Spruce budworms are normally present at low densities but occasionally undergo outbreaks with densities approaching 22,000,000 larvae/ha. These outbreaks can extend over more than 70 million hectares and last for up to fifteen years, causing a greater than 80% annual loss of foliage on fir and spruce. This foliage loss can potentially kill almost all of the trees in mature stands of balsam fir. Trees not killed by defoliation are often at risk from other insects and pathogenic diseases and standing dead trees increase the risk of fire.

While the mechanisms behind the beginning and end of outbreaks is unknown, weather may play a role in at least a portion of the budworm's range. Outbreaks in some areas follow droughts or start in stands with high concentrations of mature firs flowering after hot, dry summers. Drought stresses host trees potentially leading to a reduction in plant defense mechanisms and insects feeding on drought stressed plants often have increased growth, survival and/or reproduction rates. Drought and higher temperatures may also shift the timing of reproduction in budworms such that they may no longer affected by some of their natural parasitoid predators. Weather may also play a role in stopping outbreaks. Some outbreaks in central Canada are thought to have been halted by late spring frosts killing the new growth on trees, thus starving the budworms.

The control of some populations of eastern spruce budworm may be strongly aided by bird predators, especially some of the wood warblers. Several warbler species, including Cape May (Dendroica tigrina), Bay-breasted (*D. castanea*), Blackburnian (*D. fusca*), Tennessee (*Vermivora peregrina*) and Nashville (V. ruficapilla) are important predators of spruce budworms. Many of these species, but especially Bay-breasted Warbler, show functional responses to increases in budworms; individuals moving into the area and increasing their reproductive output in response to increases in the insect's population.

Birds consume up to 84% of budworm larvae and pupae when budworm populations are low (approximately 100,000 / ha). However, once populations exceed 1,000,000 larvae / ha, bird predation is unable to impact budworm populations. These warblers, in concert with other budworm predators (mostly predatory insects), are thought to be able to control populations of budworms under non-outbreak conditions. While there have been no recent studies on the link between bird population changes and budworm outbreaks, high warbler mortality due to inclement spring weather in Minnesota in 1907 and 1910 immediately preceded budworm outbreaks.

Climate change has the potential to impact almost every component of this system, both individually and through species interactions. For example, the spruce budworm's northern range might increase with increasing temperatures; the current species' distribution thought to be tied to whether larvae can complete development before autumn freezes begin. Increasing temperatures and drought frequency has the potential to lead to more frequent and possibly more severe outbreaks. This change would occur because of the effects that both drought and temperature have on host plants and insects. Increasing temperatures may also reduce the frequency of late spring frosts, potentially increasing the length of outbreaks in those areas.

A changing climate might also decouple budworm population cycles from those of some of its parasitoid and avian predators. Distributions of most of the warblers that feed on spruce budworms could shift significantly farther north. One set of models projects that three of the most significant predators, Tennessee, Cape May and Bay-breasted Warblers, may all be extirpated from below 50° N latitude with a doubling of CO_2. Indeed, the average latitude of occurrence of both Cape May and Bay-breasted Warbler has already shifted significantly farther north in the last 24 years. The end result of these changes may be an increase in the rate of conversion from boreal forests to other habitat types.

The combination of increasing temperatures, drought conditions and insect damage may, in turn, increase the likelihood of major fires. Two key questions are how quickly might southern boreal forests be converted to a different habitat type and what might that habitat be? As fire frequency increases, the successional pathway changes from fir/spruce to aspen/birch to grasslands. If the climate changes quickly then more southerly species may not be able to migrate into areas as fast as boreal species are lost leading to a shift to grassland or grass/shrubland. This, in turn may have impacts on both regional economies and on distributions of species using southern boreal forests. Predator-prey interactions are just one of the ecosystem services potentially impacted by climate change.

Pollination

Pollination is crucial to the reproduction of many plants and thus to the maintenance of functioning ecosystems and biodiversity. The principal mode of pollination of many plant species is by insects. Worldwide, an estimated 400 crop species are pollinated by bees and more than 30 other animal genera with a possible crop loss in some species of greater than 90% in the absence of bees. Worldwide, more than 100,000 different animal species play a role in the pollination of 250,000 types of wild pollinated plants.

Vertebrate mediated pollination may be most significant on some South Pacific islands. For example, most species of flowering plants on Samoa are pollinated by a relatively few species of animals. Plants on isolated islands may therefore be much more vulnerable to the loss of a given pollinator (or seed disperser). On many islands in the South Pacific the key pollinators and seed dispersers are flying foxes (Pteropus spp.). At least 289 plant species in Africa, Southeast Asia and the South Pacific depend, at least in part, on flying foxes. Of these, 186 plant species produce at least 448 products of value to humans.

Seed dispersal

In Costa Rican tropical deciduous forests, as many as 63% of the plants have their seeds dispersed by birds, 77% in subtropical evergreen forests and 82% in montane evergreen forests. While the number of species whose seeds are bird dispersed is higher among the plants in the understory, the seeds of many canopy plants are also bird dispersed. On Samoa, the vast majority of the seed dispersal in the dry season is mediated by flying foxes. Some plants have their seeds dispersed by one, or only a few animal species. North American whitebark pines (Pinus albicaulis) have coevolved with Clark's Nutcrackers (Nucifraga columbiana) such that the cone's structure, as well as its fruiting time, increases the likelihood of seeds being cached by nutcrackers.

In Europe, a similar mutualistic relationship exists between the Eurasian nutcracker (N. caryocatactes) and other members of the stone pine family. The propensity of Steller's Jays (Cyanocitta stelleri) to cache the seeds of piñon pines farther up wooded slopes might aid in the upward migration of this pine as climate changes. This is one of the types of systems most vulnerable to climate change, that of a keystone mutualist, where one species has a direct influence on another and significant indirect influences on many more. If a keystone mutualist disappears, an entire system is at risk, not just the paired mutualistic species. The caching of seeds by animals is a common way in which plants disperse. If animals disperse at one rate, relative to their physiology, and this rate or the animal's new range is outside of the plant's climatic tolerances then the plant species may face local extirpation.

Decomposers and soil maintainers

Species, typically invertebrates living on or below the soil surface, provide needed goods and services to human societies (e.g., mixing and aeration of soils, decomposition of organic materials). These processes contribute to the creation of fertile topsoil from organic matter and mineral components of the soil. Climate change may have already changed the diversity of the soil-decomposing invertebrate Collembola. Many species of ants, a belowground component of ecosystems, are also susceptible to climatic changes, especially droughts. Many waste products generated by society are also broken down by various wildlife species (e.g., bacteria, insects and other invertebrates, worms, etc.).

Subsistence lifestyles

People in many parts of the world are dependent upon wildlife for all or part of their daily nutritional needs. This is most pronounced in less-developed areas. For example, in the Arctic and Sub-Arctic regions of North America, local communities depend upon a mix of cash income (often derived from wildlife by hunting or trapping) and subsistence harvesting with 50% or more of the diet obtained through hunting or fishing. Wildlife are also a significant source of food and medicine for people in many temperate and tropical countries, such as Botswana and Nigeria. In many countries climate change impacts, such as reductions in wildlife populations, are likely to have the greatest impact on the lowest income groups—those having the least ability to adapt if hunting opportunities decline.

Many Indigenous Peoples use wildlife as integral parts of their cultural and religious ceremonies. For example, birds are strongly integrated into Pueblo Indian (US) communities. Birds are viewed as messengers to the gods and a connection to the spirit realm. Among Zuni Indians (US), prayer sticks, using feathers from 72 different species of birds, are used as offerings to the spirit realm. As another example, Western and Mountain Bluebirds are found on Hopi and Zuni reservations only in winter and are used as symbols of the transitions for both fall and spring. Among both the Hopi and the Zuni, bluebirds are associated with the puberty rituals surrounding the passage of a girl into a womanhood. Thus, shifts in the timing or the ranges of wildlife species could impact the cultural and religious lives of some Indigenous Peoples.

Ecosystem stability and biodiversity maintenance

Other valuable services are provided by species contributing to ecosystem stability, or to ecosystem health and productivity. Reductions in or losses of species can lead to both reduced local biodiversity, and changes in the structure and function of affected ecosystems. The most well known example of this kind of effect comes from marine systems, where the presence or absence of a starfish species was found to greatly influence the species composition of intertidal habitats. Species in terrestrial systems can also have a strong influence on the biodiversity of their ecosystems; in many cases these effects are related to their functions as pollinators or seed dispersers.

Adapting to Climate Change

Since it is unlikely that many species will be able to naturally adapt to climate change (especially abrupt climate change) then is becomes increasingly important to adapt natural resource management techniques instead. To plan and respond effectively, managers must first understand the nature of the climatic and ecological changes that are likely to occur in their region. Monitoring climate change and how ecosystems respond is important so that adjustments in management strategies can be made as needed. Numerous adaptations and combinations of approaches will emerge as experienced wildlife managers gain an understanding of the changes that are likely to occur in habitats and species in a given biome.

Unfortunately, uncertainty as to the magnitude and timing of climate change may discourage some from adapting management practices. However, moving towards what various authors have termed 'no regrets' management or 'precautionary principle' management would certainly be prudent. This approach would encourage conservation planning that is both relevant today and which models, or first principles, suggests will still be relevant in the future. This is a rapidly developing field, especially in the area of preserve design, so this list should not be considered exhaustive.

Adaptation starts with recognition that climate change is occurring. Conservation planners and managers need to become better informed about the potential consequences of climate change and variability on the resources they work with. Developing techniques that allow conservation resources to be adequately managed in the face of climate variability may ultimately prove the most beneficial way to be prepared for abrupt climate change. For example, the ability to properly manage natural resources during El Nino events potentially acting as a surrogate for the potential changes that could occur with an abrupt climate change.

Managers also need to be better prepared for climate change surprises. These surprises could come about in terms of climate or in terms of wildlife or habitat responses. For example, one type of surprise may be repeated "100-year" floods because of changes in precipitation cycles. Another example might a change in climate leading to an insect pest switching from one generation per year to two generations per year – with increased habitat damage as a result. To a certain extent, use of existing adaptive management techniques can deal with some of these issues.

One of the primary adaptations to abrupt climate change is to reduce as many of the other stresses on the resource as possible. This, in turn, may lead to an increase in the resiliency of habitats and species to the effects of climate change and variability. Accompanying the removal of other stressors is the need to maintain healthy, connected and genetically diverse populations. Isolated, small populations are often more prone to local extirpations than larger, more widespread populations. Connected populations also provide an opportunity for members of a population in one portion of a species range to move to another portion.

Finally, the more genetically diverse a population is to begin with the greater the possibility that a species might be able to naturally adapt to climate change. Thus, healthier or more robust species and habitats should be better able to adapt to climate change as an additional stress. Although these are goals managers already strive to accomplish, climate change increases their importance.

Fire is a natural part of many ecosystems. Climate change, however, could lead to more frequent fires and/or a greater probability of catastrophic fires. For instance, in areas that experience lower precipitation from climate change, reductions in soil moisture can increase drought stress on plants, making them more vulnerable to disease and pest outbreaks, thereby increasing mortality. This factor, in turn, could lead to more frequent fires or a greater probability of catastrophic fires.

Managers can use prescribed fires and other techniques to reduce fuel load and the potential for catastrophic fires. An increased in the intensity of severe weather also places wildlife at risk. Although it is not possible to avoid the disturbance itself, it may be possible to minimise the effect of the event. For example, securing water rights to maintain water levels through a drought, or having an infrastructure capable of surviving floods might minimise some impacts. Maintaining widely dispersed and viable populations of individual species also minimises the probability that localised catastrophic events will cause significant negative effects. Having multiple, widely spaced populations might also offset some of the population losses attributable to wide-spread events such as hurricanes and typhoons.

Rapidly changing climates and habitats will likely increase opportunities for invasive species to spread because of their adaptability to disturbance. Already a significant problem for native plants and wildlife, invasive species control efforts will be essential, including extensive monitoring and spot control to preclude larger impacts. Managers should look either poleward or lower in elevation to identify invasive species that may move into their areas with a warming climate. By closely watching for the arrival of the species it may be possible to control the species to keep it from becoming established.

As wildlife and habitats have declined the establishment of refuges, parks, and reserves has often been used as a conservation strategy. However, placement of conservation areas has rarely taken into account potential climate change and variability, even though the problems of climate change and conservation area placement were pointed out in the mid 1980s. In highly fragmented habitats, the placement of conservation areas on a northsouth axis may enhance movements of habitats and wildlife by providing migration corridors or stepping stones. In the absence of better data, efforts to conserve habitats for single, or small numbers of species, should be concentrated from the center of the species' range poleward.

These are areas where suitable climate is more likely to be sustained or sustained longer. Managers of existing conservation areas should start to consider climate change and variability in developing their future management plans. Specifically, this planning should include assessing the vulnerability of key taxa in the preserve and monitoring potential effects related to climate change.

Ultimately, managers may need to enhance or replace diminished or lost ecosystem services. This could mean manual seed dispersal or reintroducing pollinators for some plant species. In the case of pest outbreaks, it may mean the use of pesticides. Enhancing or replacing other services, such as contributions to nutrient cycling, ecosystem stability, and ecosystem biodiversity may be much more difficult. The loss or reduced capacity of ecosystem services may be one of the major sources of surprise from climate change and variability.

Multiple Stressors on Ecosystem Functioning

Global change and loss of biodiversity have increased concern over the cumulative impacts of anthropogenic stressors on ecosystems. We refer to a stressor as an abiotic or biotic (e.g. introduction of an alien species) variable that exceeds its range of normal variation, and adversely affects individual physiology or population performance in a statistically significant way. Natural and anthropogenic stressors can affect ecosystem functioning through changes in biodiversity, especially when ecosystem processes (e.g. primary production) are maintained by only a few species.

The accelerating rate of biological impoverishment may render ecosystems incapable of compensating for the loss of biodiversity, thereby reducing their resilience to environmental change. For example, local extinction can reduce the probability of tolerant species being present that can maintain ecosystem processes in an increasingly stressful environment.

Abiotic and biotic stressors usually do not operate independently, but rather often interact to produce combined impacts on biodiversity and ecosystem functioning. Understanding multiple stressors is particularly challenging when their combined effect cannot be predicted based on evidence from single-stressor studies -i.e. there are interactions that cause non-additive effects. Stressors are synergistic when their combined effect is larger than predicted from the sizes of the responses to each stressor alone, and antagonistic when the cumulative impact is smaller than expected.

Multiple stressors have been demonstrated to impact both aquatic and terrestrial ecosystems. For example, the collapse of Caribbean coral reefs can be attributed to an increased vulnerability to hurricanes and diseases following the loss of herbivorous fishes from overfishing. Common combinations of stressors in terrestrial ecosystems include increased eutrophication, salinisation, synthetic organic pollutants, and oxidative smog events. In northern freshwater ecosystems, three important stressors are anthropogenic acidification, drought events, and depletion of stratospheric ozone, which interact to increase acidity and exposure to DNAdamaging ultraviolet (UV) radiation by reducing concentrations of UV-attenuating dissolved organic carbon.

Given the increasing multiplicity of environmental stressors associated with global change, there is an urgent need to develop a better understanding of the interactive effects of multiple stressors on ecosystems to better predict their responses to a changing environment. Here, we present a conceptual model of how a set of species that contribute to the same ecosystem process might respond to two co-occurring chronic stressors, based on the relatedness of traits that determine the species' tolerance to each stressor.

Conceptual Model Framework

The effects of two stressors on biodiversity as a function of the correlation between the abilities of individuals or species to tolerate each stressor; we call this relationship co-tolerance. Three scenarios may be envisaged in which chronic exposure to each stressor occurs more or less simultaneously (Fig. 1).

The first scenario considers species sensitivities to stressors A and B to be unrelated (Fig. 1A). In this case, elimination of a sub-set of the species pool by either stressor will leave a sub-set of species which is "random" with respect to sensitivity to the other stressor. As a result, one stressor will eliminate a random sub-set of the species that are not eliminated by the other stressor. Here, if stressor A were already present when stressor B occurred, then ecological history in terms of initial exposure to stressor A would not strongly influence the effect of stressor B on the relative number of species that are lost. However, sensitivity of a given species to a stressor is not necessarily independent of the presence of another stressor. For example, a species may exist in the presence of either stressor A or B, but not persist in the presence of both. Lack of persistence could result from either a simple additive (i.e. total dose-dependent) or nonadditive effect, which would result from a synergistic interaction. Synergy between stressors A and B would result in a greater loss of species than is depicted graphically in our model because it would amplify the effect of one or both stressors, and thereby, expand the boundaries of the shaded areas in Fig. 1. The opposite would apply when the two stressors act antagonistically.

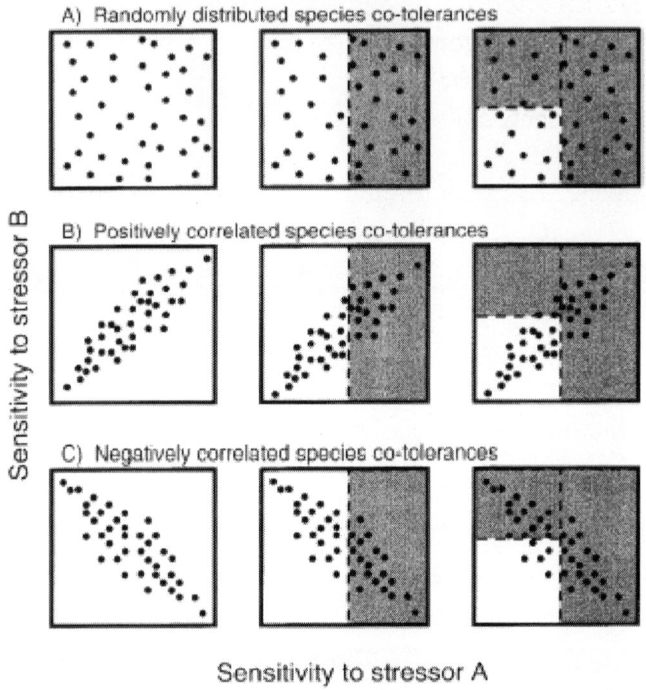

Figure 1. Hypothetical effects of two stressors on a functional group of 40 species as influenced by the relatedness of their cotolerances: A) exposure to stressors A (shaded right half) and B (shaded top half) extirpates 20 plus 10 species, resulting in a 75% loss of species when tolerances of each species are randomly distributed; B) exposure to stressor A and B leads to the extinction of 17 plus 3 species, resulting in a loss of only 50% of all species, when tolerances are positively correlated (i.e. stress-induced community tolerance); C) exposure to stressor A and B leads to the extinction of 20 plus 15 species, resulting in a 87.5% loss of the initial species pool when tolerances are negatively correlated (i.e. stress-induced community sensitivity).

The second scenario occurs when species sensitivities to stressors A and B are positively correlated (Fig. 1B). Here, either stressor eliminates certain species, but leaves more species that are tolerant of the other stressor than if species sensitivities are unrelated (Fig. 1A). Therefore, a positive correlation between species sensitivities (i.e. positive species co-tolerance) increases resistance to one stressor as a result of exposure to the other stressor. This stress-induced community tolerance, following the ecotoxicological concept of pollution-induced com-munity tolerance (PICT) where communities that have been chronically exposed to toxins are relatively less affected by exposure to new pollutants. This scenario represents a strong antagonistic interaction between stressors A and B because the former suppresses the effect of the latter, resulting in a combined impact that is less than the sum of their individual effects.

The third scenario occurs when species sensitivities to stressors A and B are negatively correlated (i.e. negative species co-tolerance) because of a trade-off between the traits that determine the ability of a species to tolerate each stressor (Fig. 1C). Here, biodiversity is likely to be severely reduced after both stressors are applied: if one stressor eliminates certain species, then the surviving tolerant species have on average a heightened sensitivity to the other stressor. This is an example of stress-induced community sensitivity, wherein exposure to one stressor synergistically increases the impact of the other stressor as determined by the relative number of species extirpated.

However, the cumulative impact of stressors A and B on the absolute number of species lost remains additive because their combined effect approximates the sum of their individual effects (Fig. 1C). Nevertheless, if synergy occurs between stressors A and B, then one or both of their individual effects would increase, which could be depicted in Fig. 1C by expanding the shaded area in our model. Therefore, a synergistic interaction between stressors coupled with stress-induced community sensitivity represents our hypothesized worst-case scenario of maximal species loss.

The location of each point in Fig. 1 representing a species should be viewed as being dynamic, and not fixed. If species sensitivities to stressors are affected by interactions with other species, then species points may move in response to a stressor. For example, the elimination of species by one stressor could release tolerant species from competition, which might better enable them to tolerate other impacts, such as the occurrence of the other stressor. Therefore, competitive release could result in a downward shift into the lower left quadrate of Fig. 1 by certain points as species become less sensitive to either one or both of the stressors. Species tolerances could also show a similar shift during chronic exposure to the stressors because of physiological acclimation, or adaptation by populations that contain intraspecific genetic variation. Conversely, species tolerance may shift towards the upper right quadrate when competition is increased as a result of release of highly competitive species from predation pressure.

Although we have interpreted the symbols in Fig. 1 as species, they could also be used to denote individuals or genotypes within a species. Species do not have fixed physiologies, and stress tolerance depends to some extent on genetic and phenotypic factors that differ among individuals. The correlations between tolerances to different stressors will also occur among individuals of the same species. If the same traits protect against different stressors, a positive correlation arises, whereas a trade-off between the traits that cause tolerance to the different stressors will lead to a negative correlation of tolerances. Therefore, points could also be represented by ellipsoids representing individual variability, stretched in the direction of the correlation between traits (Fig. 1).

Implications for Ecosystem Functioning

We can use our conceptual model to make theoretical predictions about the cumulative impact of multiple stressors on aggregated variables, such as primary production and trophic energy transfer. Evidence from a considerable body of literature suggests that the impact of a stressor is less pronounced at the functional-grouplevel than at the species-level. The resistance of aggregate functional variables (e.g. rates of ecosystem processes such as primary production) to environmental stress is usually attributed to compensatory species dynamics: tolerant species succeed competitors that are suppressed or extirpated during a perturbation, thereby maintaining overall ecosystem functioning at the same level.

Our conceptual model can be used to make qualitative predictions about the cumulative impact of multiple stressors on ecosystem functioning. The insurance hypothesis predicts that communities comprised of functional groups with many species results in ecosystem functioning being more resistant to species loss because other tolerant species in the functional group could compensate for the species that have gone extinct because of the stressor. However, environmental change often consists of several stressors that can each eliminate certain species that are tolerant of one or more of the other stressors.

The compensation effect will critically depend on the sign and magnitude of the species co-tolerance to the different stressors. The loss of species following one stressor may weaken resistance to additional stressors because species that might have been tolerant of the second stressor have been eliminated. Based on our conceptual model, therefore, we predict that ecosystem functioning is more likely to withstand an additional stressor when there is positive co-tolerance than when there is no or negative co-tolerance, because fewer species are lost.

The insurance effect would also be expected to be weakened when species that are important for maintain-ing a given ecosystem process are similarly sensitive to stressors. Consider the situation when all species in a community that play a similar role are divided into groups, rather than treated equally as in Fig. 1. For example, instead of treating phytoplankton as a homogenous group of primary producers, we can divide them into functional groups based on size, morphology, or physiological parameters. Division into size classes indicates the probability of consumption by zooplankton and subsequent transmission through the food web, while physiological categories indicate groups of species that perform unique processes such as nitrogen fixation (cyanobacteria) or biogenic opal formation.

The likely effects of multiple stressors on ecosystem functioning depend on how tolerances are distributed within and among these functional groups. When functional groups are randomly distributed in tolerance space, there is more scope for compensatory interactions than if species in functional groups have clustered tolerances. However, if species within functional groups have clustered tolerances, entire functional groups may be eliminated when there is negative co-tolerance. This situation has a real world analog: N-fixing cyanobacteria and opalproducing diatoms are extirpated from acidifying lakes in a predictable sequence. When these species go extinct, the specific ecosystem process they promote will cease.

Importantly, the impact of these losses is likely to be context-dependent; for example, the consequences for overall ecosystem functioning of Nfixer extinction depend on whether N is a limiting nutrient within an ecosystem. Nevertheless, we would expect to see changes in ecosystem functioning when functionally similar species have similar tolerances to stressors, regardless of how the functional similarity is defined.

Ecosystem functioning can also be affected by multiple stressors that alter the size structure within functional groups. There is evidence that smaller species are more often better able to withstand anthropogenic stressors. A stress-induced shift towards small species may affect ecosystem functioning in more subtle ways than the loss of entire functional groups. For example, zooplankton communities composed mainly of small species are likely to have lower per capita grazing rates and therefore lower overall grazing rates unless there are corresponding increases in density and biomass.

Moreover, because smaller zooplankton are less vulnerable to fish predation, trophic energy transfer is reduced when smaller zooplankton dominate. Although predation and other biotic processes (e.g. competition) are not typically considered to be stressors, invasion of alien species can cause biotic factors (e.g. predation pressure) to exceed their range of normal variation, and thereby, function as stressors that narrow the size structure within trophic levels such as the herbivorous zooplankton.

Common Stressors of Aquatic Ecosystems

To illustrate the applicability of our conceptual model to real ecosystems, we focus on the expected interactions associated with two common stressors of freshwater ecosystems, namely acidification and human land-use. However, our conceptual model should in principle also apply to other types of multiple stressors and ecosystems.Anthropogenically acidified lakes provide an excellent example of communities exposed to multiple stressors, allowing us to explore our conceptual model for several functional groups (Table 1). Acidification typically shifts phytoplankton communities towards larger dinoflagellates and filamentous green algae, while suppressing cyanobacteria and smaller chrysophytes and larger eukaryotic algae that are less well adapted to higher temperatures, suggesting that acidification and environmental warming have synergistic negative impacts on phytoplankton. In contrast, ultraviolet radiation has a smaller effect on phytoplankton in acidified lakes than in circumneutral lakes, suggesting that there is positive co-tolerance to pH and UV stress as a result of concomitant increases in acidity and declines in UVattenuating dissolved organic carbon in acidifying lakes.

Since acidification favours smaller zooplankton species, such as rotifers and certain copepods, we hypothesise that the effects of several other stressors on zooplankton will be reduced in acidified lakes. For example, smaller zooplankton species experience lower metabolic costs per capita than larger ones during warming events. Similarly, smaller zooplankton are less vulnerable to UV radiation, visually feeding vertebrate predators, and pesticides.

Table 1. Hypothesised antagonistic (+) and synergistic (–) effects of anthropogenic acidification (stressor A) and other major abiotic and biotic stressors (stressor B) on phytoplankton and zooplankton.

Stressor B	*Phyto-plankton*	*Zoo-plankton*	*Rationale*
Elevated temperature	–	+	Acid-sensitive cyanobacteria are thermo-philic; small acid-tolerant zooplankton (e.g. rotifers) experience lower metabolic costs than large acid-sensitive clado-cerans.
Ultraviolet radiation	+	+	Acid-tolerant species must also be UV-tolerant as they experience elevated UV expo-sure during lake acidification and loss of UVattenuating dissolved organic carbon. Small acid-tolerant rotifers show higher UV tolerance than other zooplankton.
Eutrophication	–	?	Acid-sensitive cyanobacteria are better competitors for nutrients than other groups of phyto-plankton.
Toxins	+	+	Acid-tolerant, small zooplankton species are more resistant to contaminants than large ones.
Predation	+	–	Large acid-tolerant phyto-plankton (e.g. cyanobacteria, dinoflagellates) are less edible; smaller zooplankton are more susceptible to predatory inver-tebrates like Chaoborus, while pigmented UV- and acid-tolerant zooplankton are easily detected by introduced fish.

In contrast, for certain planktonic crustaceans we would expect acidification to amplify the effects of UV radiation and predation pressure. For example, copepod species that are both acid- and UV-tolerant because of their capacity to produce photo-protective pigmentation are likely to face an increased risk of predation by visually-feeding planktivorous fish, especially if large populations of alien sportfish are introduced into clear lakes. Subsequently, smaller acid-tolerant species are more susceptible to large predatory invertebrates (e.g. *Chaoborus*), which become more abundant in lakes once they have become severely acidified and rendered fishless.

However, acidification can also amplify the effects of UV radiation on the acid-tolerant, predatory phantom midge Chaoborus because it lacks photoprotective pigmentation. Finally, adverse impacts of increased UV exposure on acidtolerant zooplankton may be offset by environmental warming because photorepair is enhanced by higher water temperatures. A number of stressors can also affect functional groups in ecosystems that are influenced by changes in land-use. For example, eutrophication and pesticides are major stressors affecting agricultural environments, while urban systems are often impacted by high salinity. Morphological and life-history traits of zooplankton adapted to eutrophic environments and elevated pesticide levels are probably positively correlated since their ability to survive either stressor depends on fast reproduction and small body size.

Accordingly, zooplankton communities stressed by pesticides are indeed dominated by rotifers and small crustacean species. In contrast, reproductive rate is inversely related to species tolerance of high salinity, probably because of energy allocation to osmoregulation. Thus, we expect that the adaptive traits for tolerance of eutrophication and pesticide exposure in lakes situated in agricultural catchments are negatively correlated with traits for salinity tolerance in urban lakes. As a consequence, functional groups that have been previously exposed to agricultural stressors may exhibit stress-induced community sensitivity and be especially sensitive to the impacts of future urban development.

Environmental Changes and Ecosystem Stability

Most scientists consider ecosystems to be dynamic rather than static systems where change and disturbance are seen as natural features of ecosystems. It is therefore a paradox to speak about "ecosystem stability". However, when breaking up "ecosystem stability" into its components and dimensions, change and dynamics becomes an integral part of ecosystem stability. In this respect, stability is rather seen as a reference against which ecosystems' responses to change is measured. As such, many properties of "ecosystem stability" has been defined of which the most recurrent are:

— resilience that is the speed with which a system returns to its equilibrium after change;

— resistance that is the tendency to remain unchanged by a disturbance;

— persistence that is the time a system lasts before it is changed.

Other important aspects of ecosystem stability relate to scale (e.g. local vs. global stability) and environmental characteristics (e.g. fragility and vulnerability) that determines how a given system reacts to disturbances. In any study or assessment of ecosystem stability, special attention has to be paid to temporal and spatial scales, against which conclusions about dynamic systems can only be made if they are properly defined. For example, fire is a natural disturbance in boreal forest ecosystems and there maintains heterogeneous landscapes both in time and space. A forest ecosystem can therefore appear to be unstable at a small spatial scale or over a short period of time, while it is considered stable at large spatial scales and long time horizons.

During the first half of the 20th century, complexity was regarded to beget ecosystem stability, a view that was built on early ecological theory. Main arguments were based on observations with a general lack of experimental support in order to exclude confounding factors. This rather simplistic view of the relationship between complexity and stability was criticised during the '70s by population ecologists who had observed in the field that animal populations were fluctuating as much in diverse as simple systems, and the exploration of more complex mathematical models that showed an inverse relationship between complexity and stability.

The complexity-stability theory has since been diversified with recent research supporting a positive relationship between complexity and stability for example in terms of functional aspects such as productivity. A general wisdom of the scientific "complexity-stability" debate has therefore been the importance of distinguishing between purely demographic (e.g. abundance and population dynamics) and functional aspects (e.g. productivity) of ecosystem stability. When studying ecosystem stability in relation to environmental change, scale and "currency" (population dynamics, productivity, etc.) are thus two important factors against which conclusions should be made. Without referring to these two factors, conclusions about ecosystem stability will be misleading.

Robert MacArthur proposed measuring the stability of an ecosystem by measuring the number of alternative pathways it contains through which energy can flow. He justified this measure by arguing that an ecosystem with many pathways, representing an abundance of species organised in a complex food web, tends to equilibrate fluctuations in population as predators will switch from less abundant to more abundant prey species, lowering population densities of the more common and allowing the density of the less common to increase.

Charles Elton gave six reasons for thinking the hypothesis to be true:

— Evidence from mathematical models suggests that those with few species are inherently unstable.
— Laboratory experiments are consistent with the mathematical models.
— Habitats on small islands are more susceptible to invasion than are those on continents.
— Less diverse habitats of cultivated or planted land are more susceptible to invasion than undisturbed habitat.
— Highly diverse tropical forest ecosystems are relatively resistant to pest invasion.
— Orchard spraying, which simplifies ecological relationships, tends to increase the likelihood of severe oscillations in pest populations.

In 1975, Daniel Goodman summarised the mounting evidence against the diversity-stability hypothesis by responding to each of Elton's arguments for it.

Models of more complex communities showed just the opposite of what Elton asserted. The more species that interact the less likely the system is to be stable.

The data suggesting vulnerability of islands to invasions by pest species may result from accidents of distribution or other special characteristics of islands.

Crops and orchard tree planted in pure stands do not represent equilibrium low diversity systems. It is difficult to find evidence that naturally low diversity communities are more susceptible to invasion than naturally high diversity communities.

The tropical biota is so diverse and complex that large fluctuations might go unnoticed. Furthermore, there is evidence that even highly diverse systems can be dramatically altered by invaders, e.g., the impact of the crown-of-thorns starfish on coral reefs.

Tilman and Downing suggest that primary productivity in more diverse plant communities is more resistant to, and recovers more fully from, a major drought.

— 207 plots of prairie grasslands differing in species richness from 1 to 26.
— Measured resistance as relative rate of community biomass change from 1986, the year before a drought, to 1988, the peak of the drought.

— Drought resistance is an increasing function of community diversity.

— Saturates at about 10-15 species.

— More diverse communities are more resistant 6 than less diverse communities, but they don't have to be very diverse.

Recent results on the relationship between bacterial species diversity and community respiration (a measure of total microbial activity) show that there are diminishing returns as the number of species in the bacterial community increases. The strong diminishing returns associated with increases in species diversity are likely to be a general feature of relationships between ecosystem processes and species richness.

In a similar experiment Tilman et al. found that plant cover is an increasing function of species richness and lower concentrations of inorganic soil nitrogen, presumably because of greater nitrogen uptake in more diverse communities.

Experiments on mycorrhizal diversity suggest that plant species composition and community structure are more sensitive to the present or absence of particular mycorrhizal associates when the diversity of mycorrhizal fungi is low. Similarly, plant species diversity, nutrient capture, and productivity are increasing functions of mycorrhizal diversity.

But there are two possible explanations for patterns like these:

— More diverse communities could increase the chances that at least one species within them is highly productive.

— More diverse communities may be able to tap resources more effectively because the differ in strategies for resource acquisition.

— Cardinale et al. perform a meta-analysis of 111 field, greenhouse, and laboratory studies that manipulated species diversity to determine its effect on abundance and biomass. They found that

— Decreasing diversity is, on average, associated with descreased abundance, decreased biomass, or both.

— The standing biomass of the richest polyculture tends to be no different from that of the most productive monoculture.

"Collectively [their] analyses suggest that the average species loss does indeed affect the functioning of a wide variety of organisms and ecosystems, but the magnitude of these effects is ultimately determined by the identity of species that are going extinct". Using a somewhat different approach

Grace et al. reach a similar conclusion: "[T]he influence of small-scale diversity on productivity in natural systems is a weak force, both in absolute terms and relative to the effects of other controls on productivity".

Europe's forest ecosystems have experienced both long-term trends and rapid environmental changes for millennia. Organisms associated with these ecosystems have evolved under these changing environmental conditions. They have responded to past natural disturbance regimes, climate changes and species interactions with adaptation or migration, others have become extinct. Recent accelerating human impact has brought in new elements affecting the ability of forest ecosystems to respond to present and future environmental changes.

For example, the practice within forestry of introducing new exotic species and moving provenances across latitudes has caused higher vulnerability of certain forest ecosystems to environmental stress factors such as pollution, extreme weather situations and insect outbreaks. Other concerns relate to nature conservation where forest reserves in Europe have become isolated islands in anthropogenic landscapes with few opportunities for the organisms to respond to environmental changes through for example migration to other more suitable areas.

Another aspect is that forestry differs from other management systems such as agriculture (that mainly is based on annual crops) in the way that it has to consider relatively long time perspectives (rotation cycles of up to 100 years or even longer, in particular in terms of nature conservation). Thus, considering the predicted global environmental change, these long time perspectives should result in reconsidering current forestry practices in order to enhance the ecosystem stability, for example by carefully revising recommendations for selecting tree species and silvicultural systems.

Environmental Changes

Environmental change has many facets as well as synonyms, which has led to misunderstandings in public and scientific debate. Additionally, the meaning of environmental change has changed many times during the past century. "Global Environmental Change" (or in short "Global Change") as used today includes not only climate change but also changes of other biophysical and anthropogenic systems, as well as it concerns local and regional scales. Environmental change is measured as an alteration of a system over a period of time, and can either be rapid changes or long-term trends (thousands or millions of years with regard to geological processes). The lack of reliable long-term data of biological, physical as well as anthropogenic components is a major problem associated with the detection and assessment of environmental change of inherently dynamic/variable systems.

Today, we are faced with a situation where current forest management goals has to be tailored against major expected environmental changes such as:

— global climate change, causing higher risk of extreme weather situations;
— land-use change, transferring soil properties and altering water fluxes;
— alteration of natural disturbance regimes, affecting ecosystem functioning;
— loss of biodiversity, with presently unknown consequences on ecosystem functioning;
— air and water pollution, leading to increased environmental stress;
— continued economic growth (energy supply incl.), acting as a pressure on natural resource management.

Consequences for Forestry

Control over one or several components of a system is inherent to the concept of management that in general strives to simplify often rather complex natural systems in order to achieve a certain goal (e.g. sustained production, maintenance of biodiversity). In extreme cases, management is expressed as monocultures with homogenous age structures (common practice in agriculture), as illustrated in Europe by plantations of Norway spruce on former cultivated land or lowland production forests of maritime pine. At the other extreme is management of nature reserves, which today is associated with the problematic of imitating or allowing natural disturbance regimes in relatively small isolated reserves, surrounded in the landscape by other land uses where these natural disturbance regimes are seen as a threat.

Landscape-based forest management responding to predicted environmental changes and that affect ecosystem stability:

— the selection of tree species and provenances (revision of current guidelines are needed and careful use of exotic species and provenances is more acute than ever);

— the choice of silvicultural systems should not be done without considerations at the landscape level (a diversification is needed);
— the consideration of ecosystem feedback mechanisms
— the awareness of practices affecting the fragmentation of and connectedness between landscape elements;

These examples are just an illustration of that future forest research needs to respond to on the one hand, the demand for reliable predictions and the development of likely environmental change scenarios, and on the other hand, the transfer of this theoretical knowledge to applied strategies to be used by forest managers and policy makers in revising current forest management practices.

Predicted climate changes will lead to direct as well as indirect effects on forest ecosystems, including altered natural disturbance regimes and consequences on biodiversity. Although, human impact has influenced landscape dynamics for several thousands of years, Europe's regions are currently going through rapid land-use changes affecting biodiversity, soils and water. Proper forest management can here be a tool for reducing soil erosion, flooding and habitat loss. Continued economic growth for an ever-increasing human population exerts a pressure on natural resource management for example through increased demand for energy and transport.

As the biophysical environment is inherently complex and involves many feedback mechanisms, effects of global environmental change are too a great extent unpredictable. This means that forests management should be adapted to a broad range of possible scenarios. Traditionally, European forestry has mainly been practised in a way that have led to resistant-based systems that are persistent to low and moderate change, while forestry with the current expected environmental change scenarios ought to strive towards more resilience-based management systems in order to better meet ongoing and future accelerating human-induced environmental changes.

References

Aber, J., Neilson, R.P., McNulty, S., Lenihan, J.M., Bachelet, D. and Drapek, R.J., (2001)."Forest processes and global environmental change: predicting the effects of individual and multiple stressors", *Bioscience*.

Bengtsson, J., Nilsson, S.G., Franc, A. and Menozzi, P., (2000). "Biodiversity, disturbances, ecosystem function and management of European forests", *Forest Ecology and Management.*

Führer, E., (2000)."Forest functions, ecosystem stability and management", *Forest Ecology and Management.*

Holling, C.S., (1973)."Resilience and stability of ecological systems", *Annual Review of Ecology and Systematics.*

Klug, J. L. and Cottingham, K. L., (2001). "Interactions among environmental drivers: community responses to changing nutrients and dissolved organic carbon", *Ecology*.

Murdoch, W.W., (1973)."Diversity, complexity, stability and pest control", *Journal of Applied Ecology*.

Tilman, D. and Downing, J.A., (1994). "Biodiversity and stability in grasslands", *Nature*.

3

Human Dimensions of Environmental Change

The history of humankind is a continuing record of interactions between people's efforts to improve their well being and the environment's ability to sustain these endeavors. Environmental constraints have led to innovations and social development, as well as social stagnation and human suffering. While the interactions throughout most of history were on a local scale, during the last decades awareness has grown so that the complexity and increasing scale of the interactions are demanding new forms of environmental management. New threats to mankind emerge, such as climate change, acid rain, ozone depletion, resource exhaustion, reduction of biodiversity, and limits to the availability of food and unpolluted fresh water. In fact, the globe is changing rapidly due to human activities, and humankind is, and will increasingly be, suffering from global environmental change.

One response from the scientific community to understand the relations between human activities and the environment is the use of modeling, that is, constructing formal descriptions of natural and societal change and their mutual interactions. Although making accurate predictions for long term future developments is inherently impossible, models can help us to show the interdependence of various activities and consequences in time and space. In that way, models can be used to communicate information and insights from the scientific community to policy makers and other stakeholders.

A number of fundamental issues in modeling the human dimensions of global environmental change are discussed. These dimensions relate to the interactions between humankind and the global ecosystem. How do human activities change our environment, how are humans affected by changes in our environment, and how does humankind respond to these changes? This contribution focuses on the behavior of humans in relation to their environment.

Although there is an increased use of modeling human dimensions of global environmental change, these issues are not well captured in current mainstream modeling practices. These issues are discussed in the context of developments in modeling human dimensions during the last 30 years and address promising developments for the future.

There are many kinds of models. A general classification is the distinction between formal mathematical models and non-formal models such as stories and cartoons. Sometimes people are role models of how one might live, such as Nelson Mandela or Claudia Schiffer. Only the formal models are dealt with here. The advantage of these formal models is that they are explicit and many of them are computer models that can be used to do repeated experiments.

Formal models are used in science very frequently, especially since the seminal work of Isaac Newton, more than 300 years ago. Formal models were mainly used to describe natural systems, such as calculating the trajectories of cannon balls or celestial bodies. Due to the success of these models, mathematical models were increasingly used in social science, especially in economics. However, human beings are not similar to cannon balls. Human beings can decide to obey traffic laws, but cannon balls cannot decide to obey the law of gravity.

The application of models from physics to social phenomena is problematic, but is still widely used. But first the field of modeling human dimensions of global environmental change is discussed.

Currently, models are widely used to describe the relations between human activities and the environment. Moreover, these models have often played a central role in setting environmental problems on the policy agenda, by exploring the consequences of alternative scenarios and by designing acceptable solutions for managing environmental problems. An example of an environmental problem for which the use of models is central is climate change. The possibility of human induced climate change is actually a policy problem that was put on the agenda after alarming model-based studies. Svante Arrhenius estimated at the end of the 19th century the consequences of a doubling of atmospheric carbon dioxide on the global mean temperature to be about 3 °C. Since the late 1950s, atmospheric carbon dioxide has been measured systematically, and currently the level is about 30% higher than pre-industrial levels. During the last 30 years, more detailed climate system models have been developed, and the results are compared with the increasing amount of (satellite) data. Still, these models are not able to describe observed climate accurately on a detailed spatial level. Besides increased efforts in modeling the climate systems, models of the human dimensions of climate change have been developed. These models were used to speculate on the possible consequences of climate change on economic growth, agricultural production and human health. Furthermore, models were developed to estimate the costs of mitigating so-called greenhouse gases.

The impacts of climate change on various social and ecological factors are based on laboratory experiments, extrapolations from field studies, historical (regional) climate-change analogues and expert judgements. The increased insights into potential impacts of human induced climate change led to the current high position of the climate change issue on the policy agenda. Policy decisions related to climate change are mainly determined by model outcomes. Many uncertainties, speculative assumptions and lack of information surround these outcomes. Thus, because the potential consequences of climate change as estimated by models are so severe, the precautionary principle is often advocated to reduce potential risks. Nevertheless, it is clear that the many uncertainties that surround the model outcomes have generated an intensive debate. Some scholars highlight the potential benefits of climate change on, for example, agricultural production, due to higher levels of atmospheric CO_2 and more suitable temperatures in Canada and Russia. Others argue that human induced climate change will not occur due to damping feedbacks of the biogeochemical cycles. Various scholars warn of positive feedbacks that may amplify climate change to catastrophic levels. The mainstream opinion on the size and impacts of climate change is represented by the reports of the Intergovernmental Panel on Climate Change (IPCC), an international scientific organization established in 1988 to assess scientific research on climate change. In 1995, the IPCC concluded that "the balance of evidence suggests a discernible human influence on the climate." Although science is not a democratic process, the assessments of the IPCC try to synthesize a mainstream picture of the climate change issue which can be used in the international policy process to formulate policies to reduce greenhouse gas emissions. Such an authorized synthesized picture on the problem is desired by policy makers, since they have difficulty in handling the many uncertainties and complexities of the climate change problem. The model-based analysis of the IPCC suggests objective assessment. The idea that models can create objective predictions of the future is widespread among stakeholders who deal with global environmental change.

Some stakeholders who doubt this objectivity do not want to use any model analysis at all. However, the traditional view that models guarantee or suggest objectivity is outdated. Model analyses, especially related to human dimensions, are highly influenced by subjective assumptions and interpretations. The challenge is to design ways to use models in such a way that they improve our understanding of reality. Improved mental models can improve decision-making. Therefore, the model-projections itself are not the most important element of model; rather it is the modeling process and the end use of models. This debate mirrors the discussion on world models during the early 1970s.

World Models

Integrated models addressing the human dimensions of global environmental change elaborate a tradition that was founded in the early 1970s by the Club of Rome. Over the past 30 years many models have been built in the tradition of system dynamics, as well as other modeling techniques. In the early days the models of the Club of Rome were criticized as being based on too few empirical data, too high an aggregation level, and too many subjective assumptions. The criticism of the earlier models still holds for many of the current modeling activities.

The models of the Club of Rome were the so-called World models. Jay Forrester developed the World 2 model during the summer of 1970 based on his system dynamics approach. According to this approach, the world can be described through a conglomeration of interacting feedback loops. The World 2 model can be considered as a first sketch of a world model, without empirically based estimation of suggested causal relations.

The collapse occurs because of non-renewable depletion. The industrial capital stock grows to a level that requires enormous input of resources, more and more capital must be used to obtain those resources, leading to less re-investment, and finally the collapse of the industrial base. Population decreases when the death rate is driven upward by a lack of food and health services. If the resource base is assumed to be much larger, collapse happens only a few years later due to food shortages and/or high pollution levels. Although it was recognized that there are various shortcomings in the models, Meadows and his colleagues argued that the model behavior was fundamental and general, and sufficiently developed to be of some use to decision-makers.

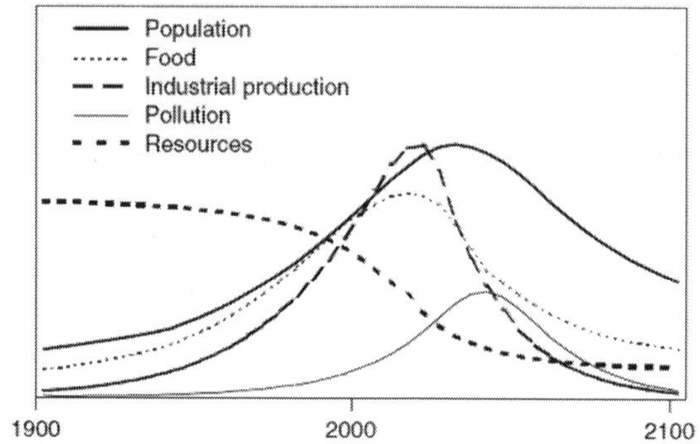

Figure 1. Standard Run of World 3

Managing Uncertainty, Complexity and Incomplete Information

Numerous scholars have criticized the World-model studies. Economist WD Nordhaus classified the World 2 study as misleading since it was not empirically tested enough and ignored mainstream economic insights. Other scholars concluded that due to the scarcity of relevant empirical information, relationships in the world models are subjective. Given the uncertainties, other sets of equally plausible assumptions can lead to a completely different picture. In fact, the outcomes of the models are largely the mental models of the model builders.

Some of the criticism was misplaced. The model outputs were interpreted as predictions, not merely as scenarios, i.e., 'what if' futures. So, when the predictions did not come true, scientists were blamed for inaccurate doomsday forecasts. Actually, the *Limits to Growth* scenarios had a profound effect on the public and government. Scientific criticism on the World models of 30 years ago mainly concentrate on two topics. First, subjective assumptions had to be made about model relations and parameters due to incomplete knowledge or even ignorance. This was especially important for the linkages between subsystems. Examples were the effects of pollution on health, the interaction between demographic and economic dynamics, the role of technological innovations in resource availability, and the relations between material and energy inputs and economic output. Second, the complexity of the underlying subsystems and the linkages made it questionable whether a high aggregation level can lead to meaningful and relevant interpretations and results.

These problems of subjectivity and aggregated relations are still relevant for the current use of models. Although the use of models has increased, we still have no satisfactory solutions of how to manage incomplete information and insights, large uncertainties and different scales. This is illustrated here by the characteristics of the current generation of global models.

Integrated Assessment Models of Global Change

Global modeling re-emerged during the early 1990s as integrated assessment modeling, mainly because of the importance of the global climate change problem. Integrated assessment models try to describe quantitatively as much as possible of the cause –effect relationships and the cross-linkages and interactions between the elements of the world system. Integrated assessment models are usually composed from meta-models of various subsystems. A meta-model is a simplified, condensed version of a more complicated and detailed model, which provides approximately the same behavior as the expert model from which it is extracted.

Integrated assessment models are one of the tools in the toolkit of integrated assessment. Other tools are policy exercises, dialogues between science and policy people, data analysis, scenario analysis and expert models. Integrated assessment is therefore a broader approach aimed at helping prioritize policy-making and research activities and giving insights into uncertainties and missing links of knowledge. It is used in a process whereby knowledge from a variety of scientific disciplines is combined, interpreted and communicated, with various stakeholders such as scientists, policy makers and non-government organizations involved.

The integrated models that are used describe the whole cause–effect chain from economics, energy production, emissions, land use changes, to changes in the biogeo-chemical cycles, the climate system, and impacts of climate change on human activities and the environment. There is no single approach to capturing the complexity of the system as a whole. In general, two types of approaches can be considered. The first approach, *process-oriented modeling*, is rooted in natural science and simulates the consequences of economic development on energy production, land use changes, biogeochemical cycles, climate system and impacts of climate change. The models are often detailed at the spatial and temporal level. The other approach, *cost–benefit modeling*, is rooted in economics and maximizes discounted long-term welfare. Models using this second approach describe the physical consequences in less detail, but express the impacts of climate change and the efforts to reduce emissions in monetary units in order to derive an optimal response by balancing costs and benefits.

Both approaches are confronted with the same dilemmas as the World models of the early 1970s. Although much more empirical information is available, many components in models are surrounded by large uncertainties. Therefore, subjective assumptions have to be made. Also the complexity of scales remains an unsolved issue.

Besides these problems, most integrated assessment models have limited abilities to capture the broad human dimensions of climate change. The models are suitable for generating projections of economic output and the costs of climate change, but they are limited in many ways. Simple macro relations are assumed between economic activities and physical processes, but empirical insights are conflicting. Current (economic) models focus on utility of consumption as the driving force of human behavior, although it is questionable how this is related to the quality of life. There is limited insight into the physical dimension of satisfying human needs, but this understanding is needed before something plausible can be said about decoupling economic development and environmental pressure. Current integrated models include only the free market as an institution, although there are many other forms of economic organization and institutions.

One of the reasons for the limitations of the current generation of integrated models is the use of a rather mechanistic modeling paradigm, which is not able to include novelty, evolution and surprise. The role of modeling paradigms is now discussed in more detail.

Modelling Paradigms

A lot of controversies among modeling studies are caused by the difference in modeling approach that is adopted by the various scientists. Each modeling approach, or in a broader context, each modeling paradigm, involves its own set of theories, concepts, mathematical techniques, and accepted procedures for constructing and testing models. We can distinguish between deterministic and stochastic models, simulation and optimization models, reductionis-tic and integrated models, linear and non-linear models, one-agent and multi-agent models, and so on. Instead of discussing all kinds of possible paradigms, the difference between the reductionistic Newtonian approach and the complex adaptive systems approach is examined. This illustrates the transition in science that is currently occurring and explains fundamental differences on how to use models for assessing the future.

Mathematical modeling has long been influenced by physical science, which has developed a mechanistic, reversible, reductionistic and equilibrium-based explanation of the world. This proved to be very successful in calculating trajectories of moving objects (e.g., cannon balls) and predicting the positions of celestial bodies. The work of Isaac Newton, culminating in the *Principia Mathemat-ica Philophiae Naturalis* in the late 17th century, was, and still is, very influential. The associated rational and mathematical way of describing the world around us was also applied in the social sciences, economics and biology. Despite the fact that later developments in the natural sciences seriously constrained the applicability of the mechanistic paradigm, its relative simplicity had a great appeal to scientists from various disciplines working with models. However, despite the widespread use of this approach, the mechanistic paradigm is increasingly criticized. The foundations of the mechanistic view: reversibility, reduc-tionistic, equilibrium-based and controllable experiments, have faded away in the light of a number of new scientific insights.

First, the discovery of the Second Law of Thermodynamics brought down the notion of reversibility. The Second Law states that the entropy of a closed system is increasing. This means that heat flows from hot to cold, so that less useful energy remains. One of the consequences of the Second Law is the irreversibility of system behavior and the single direction of time. Changes within systems cannot reverse back just like that (irreversible). This is in contrast to many mechanistic models, in which time can easily be reversed to calculate previous conditions.

Second, the equilibrium view of species was brought down by Charles Darwin's book on the origin of species during the middle of the 19th century. The static concept of unchanging species was replaced by a dynamic concept of evolution by natural selection and adaptation of species, thereby fundamentally changing our view of nature. Natural systems are in continuous disequilibrium, being interdependent and constantly adapting to changing circumstances.

Third, the theories of quantum mechanics have confronted us with a fundamental uncertainty regarding knowledge about systems, especially on the level of atoms and particles. The uncertainty principle of Heisenberg is well known, stating that it is impossible to simultaneously measure the position in space and momentum (mass times velocity) of any particle. The statement by Laplace in the early 19th century that if every position of every atom was known, the future might be predicted exactly, became therefore a lost illusion. Moreover, the notion of fundamental uncertainty implied that fully controlled experiments are strictly speaking not possible.

Notwithstanding the fact that these developments in the natural sciences changed our perception of the world, (mathematical) models are still mainly based on a mechanistic view of systems. For example, current mainstream economics is based on its success during the period after the Second World War, which was characterized by stable economic growth. Technology could seemingly handle any difficulty that came along. Affluence was seen as growing and permanent, and the standard of living, it was believed, would continue to improve for individuals and generations. The world economy was a place of simple equilibria and linear responses. Cost–benefit analysis, optimization and econometric models seemed to be quite appropriate.

Since the 1970s, various events have made us aware of the non-linearity of economic systems. The oil crisis ended unlimited economic growth, the Berlin wall collapsed in 1989, the Asian financial crisis at the end of the 1990s, the various crises on the stock markets, and so on. The economic system seems to be characterized by unstable states, non-linear responses and unpredictability. Still, mainstream economics uses its successful tools of the early days, probably because of their analytical power; certainly not because of their ability to explain reality. Increasing numbers of economists are trying new tools to explain the observed behavior of economic systems. This has led to the study of non-linear dynamics and evolutionary processes as emerging fields in economics although not always accepted by mainstream economists. This new type of economics studies the formation of patterns, evolution of economic systems, endogenous technologies, and so on. Furthermore, next to analytical tools, this new field of economics uses computers as a kind of laboratory to test hypotheses.

The emergence of new ways of studying the world around us has also emerged in other disciplines where studies are made of the origins of order, self-organization, emergence of structures, adaptation of agents in a changing environment, and many more new ideas. The general focus of this new modeling paradigm is to study how systems change and organize their components to adapt themselves to the problems posed by their surroundings. Examples of these systems are economies, ecosystems, immune and nervous systems, organisms and societies.

Although various scholars long ago discussed the importance of studying the evolution of systems, the rapid developments of the computer have provided scientists with a new tool in recent decades, which can be used to investigate evolution, self-organization, interactions between agents and emergence of structures on a macro scale by simple local rules. These systems can be grouped under the common name complex adaptive systems. They are studied by a number of new computation-based modeling tools, including genetic algorithms, cellular automata, neural networks, and artificial life forms.

The characteristics of these new types of tools are illustrated using one of these tools: genetic algorithms. *Genetic algorithms* have been developed to simulate the process of natural selection by considering a population of agents producing offspring who are similar, but not identical, to their parents. This process depends on three genetic operators: selection, crossover, and mutation. *Selection* means that the genetic algorithm selects *n* copies of the strings (genetic code) in the population by a random process that favors the most fit. Subsequently, these copies are probabilistically paired in a mating process whereby each pair produces two offspring by means of crossover and mutation. *Crossover* means that two offspring are created with a certain probability that the genetic information is crossed over; otherwise, the offspring are identical to the parents. In the case of crossover, the parent strings of genetic information are split randomly and are swapped to shape two new strings. Each element of the genetic information has a small probability of being altered. This *mutation* is independent of what happens with the other genetic information. Due to their adaptive characteristics, genetic algorithms are powerful tools for improving and finding good solutions even in complex changing environments. Moreover, genetic algorithms are based on irreversible changes, stochastic processes and evolution.

Genetic algorithms are one of the computational tools used for developing models to study complex adaptive systems, which emerged as a new field in science. Since the late 1980s, the Santa Fe Institute has provided a prime focus for exploring and deepening the insights from complex adaptive system studies and provided new opportunities for transdisciplinary studies. It is expected that the study of complex adaptive systems will become an important field in transdisciplinary environmental science. The Resilience Alliance is one of the first international communities to study environmental science from this new perspective. An overview of their work is given in Gunder-son and Holling.

Validity of Models

An important difference between the two modeling paradigms is how to use models, and how to validate models. From the viewpoint of mechanistic explanations of reality models are valid when predictions generated by the model are not rejected. An example of this Popperian type of validation is the development of an econometric model. The timeseries of data is split into two. One part is used to estimate the model and to generate predictions, and the second part to test the predictions of the constructed model. This empirical validation leads to problems according to the complex adaptive systems paradigm, which focuses on evolving structures and is often based on qualitative insights. Since small differences in initial values can lead to large output differences, models of complex adaptive systems are not suitable to generate predictions. In fact, this problem goes back to the classical three-body problem (e.g., the system involving Earth–Moon–Sun), which has been studied extensively with a variety of methods beginning with Newton. Despite the large number of studies, no complete analytic solution in closed form has been found. Furthermore, for some initial conditions this system is found to produce chaotic unpredictable behavior. This example shows that when interactions among agents are taken into account, the Newtonian approach does not work. Therefore, validation in this line of modeling is based on expert judgements and relations with theoretical insights. This type of validation can be called conceptual validation. In fact, validation is not a test, but a subjective process.

The difference between opinions on how to measure the quality of models leads also to differences in how to use models. According to the Newtonian modeling paradigm, a model that is tested empirically and is validated can be used to make accurate and objective predictions. However, according to the complex adaptive system perspective, models can only be used to study qualitative structures, and should be used interactively to be able to exchange insights and stimulate discussion on uncertainties. In the physical sciences, one no longer refers to model validation, but rather to model performance, which is provided by statistical measures such as the root-mean-square differences between model predictions and observed values.

If one wants to describe the orbits of planets, a mechanistic model is a perfect tool. Mechanistic models are even suitable tools when one wants to send men to the Moon. But in situations of many uncertainties and surprises, such tools will not work. In the case of living beings, mechanistic approaches are of limited explanatory power. From this new perspective, we look at how to design and use models for exploring the future. But first, one of the main seeds of uncertainty and subjective assumptions is discussed: stability characteristics of systems.

Myths of Systems

Different perceptions of reality can be visualized by different myths of stability, that is the perception of how systems function. According to the equilibrium myth, systems are in equilibrium. External effects can push the system briefly out of equilibrium, but it automatically returns back to the previous equilibrium situation. This myth corresponds very well with the Newtonian-modeling paradigm. Not only the natural system is considered to be in a natural equilibrium, but also the economic system of supply and demand is in equilibrium due to control of the invisible hand. This metaphor from economist Adam Smith, proposed at the end of the 18th century, was a powerful explanation of micro-behavior in order to describe an elegant mechanical description of the macro-level of economic behavior.

The myth of stability can be represented graphically as a ball at the bottom of a valley. Perturbations only temporarily knock the ball away from the bottom of the valley. An implicit assumption of this myth is that systems have the capacity to damp all kind of disturbances.

An alternative myth is the obverse, namely the myth of instability. Systems are assumed to be very sensitive to disturbances. Every disturbance can lead to a catastrophe. Applied to environmental issues, the myth of instability explains why some people argue that human activities should not be allowed to disturb the natural system. Any degree of pollution or degradation of extraction can lead to a collapse of the system. This myth can be visualized by a ball on a peak. Any perturbation can lead the ball to roll down the slope.

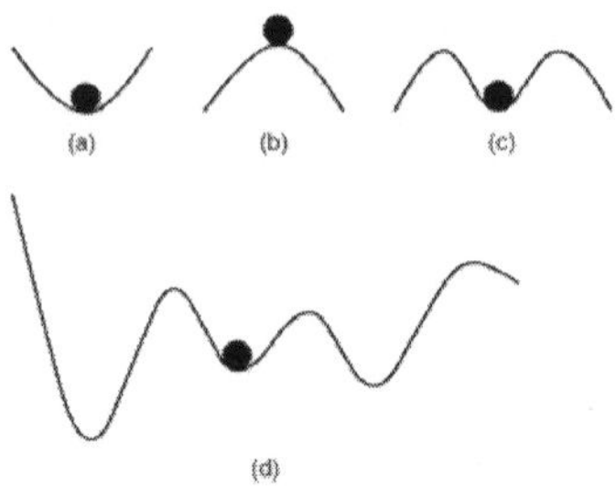

Figure 2. Myths of nature: (a) nature is stable; (b) nature is unstable; (c) nature is stable within limits; (d) nature has different stability domains

A third myth is in between the myths of stability and instability, namely a system is assumed to be stable within limits. When the system is managed well, the system can absorb small perturbations. This myth can be visualized as a ball in the valley between two peaks. A more advanced framework is to consider multiple stable states. As depicted, this myth can be represented as a number of peaks and valleys. The ball is resting in a valley and is able to absorb a certain degree of disturbance. However, an extreme disturbance can push the ball over a peak such that it will rest in another valley, an alternative equilibrium state. Examples of these multiple states are lakes, which can flip from an oligotrophic to a eutrophic state due to inputs of phosphates, and rangelands that flip from a productive cattle-grazing system into a less productive rangeland dominated by woody vegetation, triggered by variability in rainfall.

Myth of Resilience

The myth of resilience does not only consider the balls moving up and down the peaks and valleys, but also considers possible movements of the peaks and valleys themselves. In this evolutionary picture stability domains can shrink, and disturbances that previously could be absorbed no longer can be. This view has important implications for managing systems. From the perspective of the previous myth of systems, systems could be known perfectly. Surprises may lead to changes of management, because the balls moved in another valley, but in principle, management was simply a matter of controlling the system. From the perspective of an evolving landscape, one has to manage a system in the face of fundamental uncertainties over the functioning of the system. One continually observes the system in order to respond adequately. Moreover, small human-induced perturbations are supported in order to learn from the system. This type of management is called *adaptive management.*

Holling has proposed a framework for resilience to explain the transitions in behavior of the system. He distinguishes four basic functions common to all complex systems, and a spiraling evolutionary path through them. This evolutionary cycle can be used to explain transitions in social systems, as well as in ecosystems. The central idea is that the four-phase adaptive cycle emphasizes a loop from conservation to two phases of destruction and reorganization in which innovation and chance assume dominant roles. The reorganization phase occurs when a rare and unexpected intervention or event shapes a new future. In this state, the system is most likely to be transformed by innovation, and agents have the greatest potential to influence the future of the system. When the agents do not react properly to changes in the system, it can flip into a new kind of system.

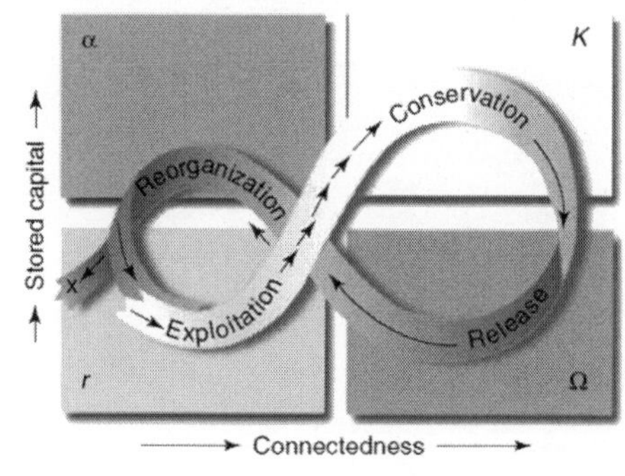

Figure 3. The dynamics of a system as it is dominated by each of four processes: rapid growth (*r*), conservation *(K),* release *(S2),* and reorganization (a). The arrows indicate the speed of the cycle. The short, closely spaced arrows indicate a slow change, while the long arrows indicate rapid change. The cycle reflects systemic change in the amount of accumulated capital (nutrients, resources) stored by the dominant structuring process in each phase, and the degree of connectedness within the system. The exit from the cycle at the left of the figure indicates the time at which a systemic reorganization into a less or more productive and organized system is most likely to occur.

As mentioned before, the landscape is changing. This has much to do with different speeds of change in the various scales of systems. For example, due to phosphorus accumulation in sediments of lakes, recycling of phosphorus from these sediments can lead to surprises. The slow variable, the sediment, can reduce the capability of the lake to absorb an external disturbance. The equilibrium levels of concentration of phosphorus in the water, the fast variable, can therefore change due to changes in the slow variable.

According to the myth of resilience, problems could be caused by local human influences that slowly accumulate to trigger sudden abrupt changes that may affect the vitality of societies. There are counteractive forces that give ecological systems the resilience and adaptability to deal with considerable change, and that provide people with the capability to innovate and create. However, nature, people, and economies are suddenly now coevolved on a planetary scale, each affecting the others in such novel ways, and on such a large scale, that large surprises may overwhelm the adaptive and innovative capabilities of people. One challenge of sustainable development is, therefore, how to stimulate coevolution of human activities and environmental change.

Different Perspectives on the Problem of Climate Change

The discussion of myths shows various possible subjective interpretations of reality. The importance of different perspectives for modeling the human dimensions of global environmental change is now illustrated. There are various concepts designed to classify different worldviews. Like the case of modeling paradigms, there is no true classification of worldviews. A contribution, which gives a general description of perspectives on natural and human systems and social relations, has been made by Michael Thompson and his colleagues in their cultural theory. This theory has been used to classify different types of institutional designs in relation to global environmental change. The cultural theory has been an inspiration for implementing worldviews in formal models, because it includes perspectives on human as well as natural systems, it claims generality, and its explanation of perspectives' rationalities is deterministic. The cultural theory combines anthropological insights with ecological insights, resulting in different cultural types.

The three main worldviews in the Cultural Theory are the following types:

1. Individualists assume that nature provides an abundance of resources, and remains stable under human interventions. A responsive management style is advocated.
2. Egalitarians assume that nature is highly unstable, and a small human intervention may lead to complete collapse. A preventive management style is preferred.
3. Hierarchists assume that nature is stable in most circumstances but can collapse if it crosses the limits of its capacity. Therefore, control is advocated as a management style.

As discussed earlier, human induced climate change is a topic surrounded with many uncertainties. It is, therefore, an excellent example to illustrate how worldviews can be quantified to simulate alternative futures based on different perceptions of reality. Such an analysis has been made by Janssen and de Vries who developed a simple integrated ecological economic model for which they implemented three versions based on alternative assumptions on climate sensitivity, technological change, mitigation costs, and damage costs due to climate change.

Egalitarians, for example, assume high climate sensitivity, high damage costs, low technological development, and low mitigation costs. For management styles, they assume different strategies for investments and reductions of emissions of carbon dioxide. By contrast, the individualist, for example, assumes a strategy that maximizes economic growth, and emissions are reduced only if a certain threshold of economic damage is exceeded. The hierarchist tries to balance economic growth and climate change by assuming the IPCC estimate of climate sensitivity.

Suppose that all of the agents in a model world share one of the three extreme worldviews. If agents are assumed to have perfect knowledge of their world, their utopia can be simulated. If their worldview is incorrect and they still apply their preferred management style, their dystopia can be simulated. However, if the individualistic world-view manages a world that actually operates according to the egalitarian worldview, emissions increase until climate change causes such an economic disaster that emissions are reduced by the collapse of the economy.

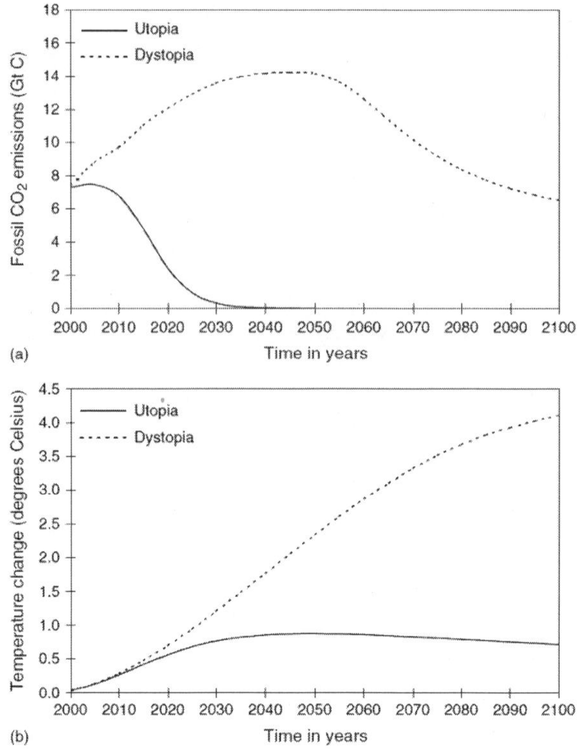

Figure 3. Expected carbon dioxide emissions (a) and temperature increase according to the egalitarian utopia and a possible dystopia (b) (individualistic management style in an unstable global system)

By introducing a population of agents with heterogeneous worldviews, a complex adaptive system is produced. It is assumed that the better an agent worldview explains the world's observed behavior, the greater is the chance that an agent will not change its worldview. On aggregate, there is a trend towards changing to the worldview that explains the observations in the most likely way. Suppose that reality is one of the three possible worlds, and an agent obtains information over time that causes him to change (or not) his perspective on the climate change problem. Three sets of projections are derived in which agents adapt to climate change. Prior to the year 2040, the observed climate change does not lead to domination of one of the worldviews. After 2040, the climate signal becomes clear enough that one of the worldviews begins to dominate.

In the event of the world functioning according to the egalitarian worldview, the emissions growth stabilizes on average in the coming decades and decreases to a level below half the present amount of emissions. However, this reduction cannot avoid a global mean temperature increase of about 2.5 °C in the 21st century.

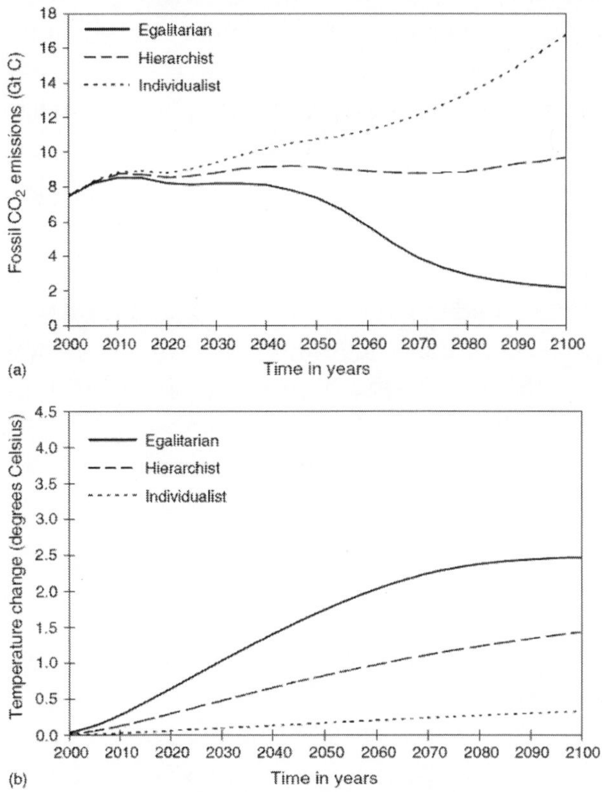

Figure 4. Expected carbon dioxide emissions and (b) temperature increase according to different views on the functioning of the global system, and where the worldviews of agents change in time

The explicit inclusion of subjective perceptions of reality has led to a rich variety of possible futures. This will not simplify decision-making, but can improve decision-making since a large set of plausible scenarios is presented. This type of modeling is currently applied in different areas of global and regional environmental change. Each system in which it is not certain that it will remain stable under all kinds of disturbances, can in principle be studied by explicitly assuming different perceptions of reality. It is the expected that this type of modeling will be especially interesting for simulation of institutional dynamics.

Macro-level Dynamics

Although there are often uncertainties about the relations, global models simulate macro-level dynamics. Here, we focus on the most important macro-level drivers of global environmental change, which are population size, economic development and technological change. In the early 1970s, Paul Ehrlich and John Holden described the environmental impact of society by the well-known IPAT equation: *I* D *P* ð *A* ð *T*. Here environmental impact equals the product of population size, the degree of affluence per person and the environmental impact from the technology used to produce one unit of affluence.

The coming sections discuss three factors that are closely related to the IPAT equation: population growth, material and energy consumption of economies, and technology development. For each factor, empirical information on the macro-scale provides information on possible developments in the long term. However, extrapolating historically derived macro-level relations involves subjective assumptions. Therefore, it is important to explore the consequences of these subjective aggregated relations and to improve understanding of the empirically derived macro-level information.

Demographics and Human Health

Population growth is one of the main causes of global environmental change. The continuing population growth can be explained by changes in demographics and health. The so-called demographic and health transition describes how populations can go through typical demographic and health stages when they change from living in pre-industrial conditions to having a mortality pattern that is found in the post-modern societies.

The shapes of the demographic transition curves are well known, but are in fact a hypothesis based on cross-national and longitudinal studies. In developed countries the demographic transition has reached the final stage. But how will this development of fertility and mortality figures continue in the future? Will the developing countries follow the same transition, and at what rate? Most population projections are based on the assumption that all countries will go through the demographic transition leading to a leveling off of population growth during the 21st century. Observed transitions in demographics are the result of changes in individual behavior, technology and norms, improvement of health care, use of contraceptives, age of marriage, literacy, position of woman, regulation of abortion, etc. Assumptions about the health transition in various regions of the world are even more difficult since various diseases are related to global environmental change (skin cancer due to stratospheric ozone depletion), behavior (lung cancer from smoking), emerging new diseases (acquired immune deficiency syndrome – AIDS) and even some re-emerging old diseases due to the development of new resistances (malaria). It is therefore difficult and subjective to project health conditions of our descendants.

Material and Energy Consumption of Economies

An economic system can be viewed as a living system, consuming material and energy inputs, processing them into usable forms, and eliminating the wastes. The metabolism of economies has changed significantly during the last two centuries. A world economy has emerged that produces agricultural, and industrial products and services in large volumes, and transports them all over the globe. Agricultural production has increased due to more and more intensive use of land. Productivity has improved due to biological innovations in the form of new crops, new agricultural practices, mechanization, and increasing synthetic inputs. The same holds true for industrial production, which is increased due to the increasing consumption of energy and materials. Currently, the service sector is becoming increasingly important, stimulated by increased personal mobility and more individualistic lifestyles.

These economic developments have led to an increase of material and energy use in absolute figures, but also in *per capita* figures, and have led to all kinds of disturbances of biogeochemical cycles. These disturbances have led to environmental change on various scales. The debate whether the environment is able to sustain economic development goes back to Thomas Malthus at the end of the 18th century, who argued that food production could not be increased quickly enough to keep pace with the growing population. But due to a faster increase of agricultural productivity than expected, a decrease in birth rates, and the growing import of food, Malthus' homeland did not collapse. Since the early 1970s, stimulated by the *Limits to Growth* report of the Club of Rome, the debate on economic growth and the environment has emerged again.

In the last 30 years, there have been attempts to improve the physical reality of economic models by including mass balance conditions, and laws from thermodynamics. However, there is still no clear theory on the relation between economic development and the environment. In the 1990s a lot of empirically based studies were published on the relation of environmental pressure of an economic system and the average income.It consists of three phases:

1. initially income growth parallels progressively increasing environmental pressure;
2. next, further income growth leads to increasing environmental pressure until it reaches a maximum;
3. further income growth leads to a reduction of environmental pressure.

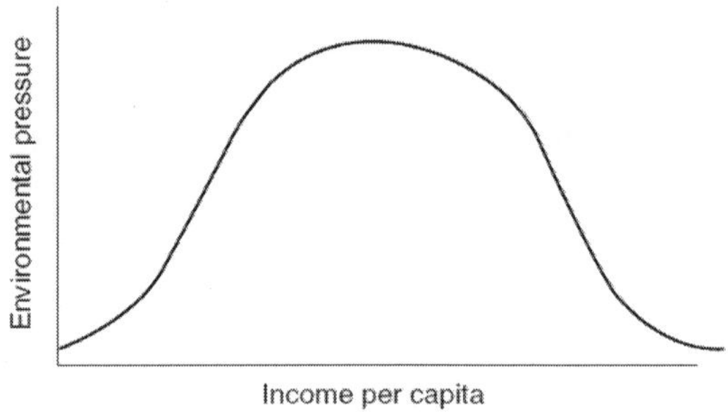

Figure 5. Environmental Kuznets curve

An explanation for this pattern is that at higher income levels, individuals will attach more value to environmental quality; this means more income spent on less damaging consumption, as well as more democratic support for stringent environmental policies. An important implication of the environmental Kuznets curve is that growth by itself would be able to solve environmental problems. However, the empirical support for this hypothesis is weak and mainly based on cross-sectional analysis. It is not clear whether the curve differ for different types of environmental pressures, what is the influence of policy measures and technological change, and whether observed trends in the past will continue in the future. Some studies suggest that delinking of environmental pressure and economic output has only been a temporary phenomenon caused by efficiency and technology improvements after the oil crises of the 1970s. The cheap energy prices of the 1990s led to a relinking of environmental pressure and economic output. Despite the many uncertainties, relationships between physical and financial dimensions of economic systems like the environmental Kuznets curve are used to explore future material and energy consumption.

TECHNOLOGICAL CHANGE

Technology development is the source and remedy of environmental change. This is called the paradox of technology and the environment. It is the source, because it creates the ability of societies to mobilize more materials and energy, and because it creates new materials and substances with direct environmental impacts. On the other hand, technology can also be the remedy because it increases productivity and efficiency of economic activities and invents specific technologies to prevent pollution.

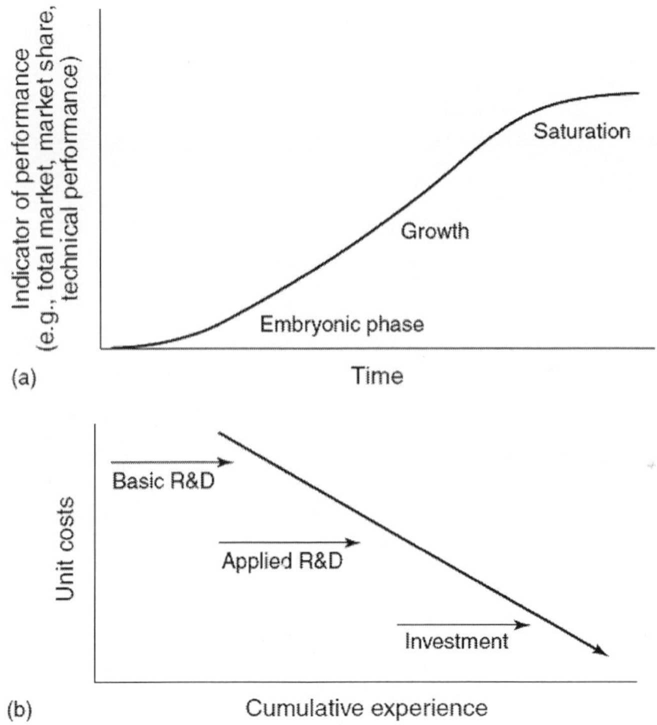

Figure 6. A stylized technology life cycle; (b) a stylized learning curve, where axes scales are logarithmic. In the beginning costs decrease due to basic R&D. When the potential of the technology is demonstrated, applied R&D investments reduce the costs further until a level is reached for which costs are competitive.

It is therefore important to understand the dynamics of technological change. The question is how a new technology is adopted at a large social and spatial scale. This can be visualized by the stylized technology life cycle. In the beginning, a new technology is imperfect, and various possible designs are explored. The market effect is small, but the increase that occurs during the growth phase is characterized by increasing standardization and falling costs. Finally, the growth rate slows down as the market becomes saturated. In this phase the market is in the hands of a few suppliers. Technology life cycle is related to the developments of costs. There is a lot of empirical evidence that the decrease of a technology investment cost is related to the accumulated investments in the particular technology. This learning-by-doing can be formulated by the technology learning curve. The costs (on a logarithmic scale) decrease linearly with an increase in cumulative experience. The main uncertainty of technology in the early phase of development is the slope at which the costs decrease.

In mainstream economic models, technology is often included as an exogenous variable. Such an assumption leads to wait-and-see policies because investments are delayed until clean technology has become available at suitable cost levels. Such a policy differs from insights derived from the learning curve, which suggest stimulation of investments in clean technology allowing costs to decrease. Insights into technology dynamics show stylized facts derived from empirical studies. This provides tools for modeling technology development. The main question is how does an individual technology penetrate the market, and how can governmental policy stimulate the diffusion of green technology? Many decisions of individuals influence this evolution.

Understanding the Emergence of Stylized Facts

The observed stylized facts, the observed macro relations, are the result of decisions of many agents. They are emergent properties of a complex adaptive system. To improve our understanding of these observed stylized facts, we should develop tools to understand them. One way is to use models that are able to simulate these emergent properties from the bottom up. Like the cartoon, understanding of macro level phenomena can only be derived by studying the behavior of micro level agents. In the case of a school of fish, multi-agent studies have shown that flocking behaviour can be simulated by using three simple rules for each agent:

- separation: steer to avoid crowding local flockmates;
- alignment: steer towards the average heading of local flockmates;
- cohesion: steer to move toward the average position of local flockmates.

Behavior of human agents is much more difficult to capture by simple rules. Two types of human behavior can be distinguished: individual behavior, and behavior of groups and institutions.

Individual Behavior

Since theory development in social science is rather fragmented, various models of human behavior exist. In fact, different disciplines study only particular aspects of human behavior. One central element, or better, one stylized problem, is found in most models of individual behavior. Humans are assumed to maximize their well being given budget constraints. These budget constraints mainly relate to income and time. Within economics and psychology, different variations exist on this stylized problem. In the formal approach of conventional economic theory, this means that the rational actor, the *Homo economicus*, maximizes its own well being. This *Homo economicus* is assumed to have perfect knowledge of the system in order to find the global optimum. In the case of uncertainties, the probability distributions are perfectly known. These assumptions make it possible to formalize human behavior in an unequivocal way. A drawback is the existence of much empirical evidence that real humans do not behave in this way. For example, experiments show that humans discount the near future at a higher rate than the distant future, experience well-being with relative changes instead of absolute levels, and are about twice as averse to taking losses as to enjoying an equal level of gains.

Some economists, like Nobel Laureate Herbert Simon, argue that humans are rationality bounded. First, no person can ever assemble all the information required for an optimal decision. Second, even if one could, decisions are usually so complex that no simple algorithm exists for evaluating all possible options. Third, a person's own decisions depend on the decisions of other persons. Simon argues therefore that humans satisfice instead of maximize their well being. A consequence of relaxing the assumption of maximizing behavior is the large set of possible rules that describe behavior. The formalization is not unequivocal anymore. Thanks to the development of new simulation techniques such as cellular automata, genetic algorithms and neural networks, and the widespread availability of personal computers, social scientists are exploring new ways of modeling human behavior.

Psychologists and other social scientists include different indicators than are usually used in economics. Psychologists focus on satisfaction of various types of needs such as understanding, freedom, identity and leisure. Furthermore, all humans are different. They differ in their abilities, mental models, preferences and opportunities. In contrast with *Homo Economicus*, decisions depend on social interactions, as well as on individual considerations. In fact, humans are assumed to perform different cognitive decision processes in different situations.

The mental models of humans are important elements in cognitive processing. Differences in perceptions of reality can result in different behavioral patterns. Although humans can learn, they will never have the perfect knowledge that is needed to maximize well being as performed by *Homo Economicus*. This picture of human behavior is in line with the complex adaptive system-modeling paradigm. Since humans cannot perfectly control their own situation, they design institutions to regulate human interactions, as well as interactions between society and the environment.

Institutions and Collective Actions

Institutions contain formal constraints (rules, laws, constitutions), informal constraints (norms of behavior), and their enforcement characteristics. They shape human interactions and the way societies evolve through time, and can also be important to regulate human activities in relation to ecological conditions. A stylized problem that is generally used to study institutions is the so-called *commons dilemma*, widely discussed as a result of the well-known analysis of Garrett Hardin on the *Tragedy of the Commons*. According to this analysis, the commons tend to be overharvested since each agent harvests to the point where private costs equal the benefits, whereas harvesting imposes additional social costs on the rest of the community. However, historical analyses of common resource properties have found many examples where the tragedy did not happen. Communities often had ways of self-organizing to prevent overex-ploitation of the commons, also known as closed commons. Success of self-organization depends heavily on the characteristics of property-right systems. The tragedy of the commons is an example of open access, where everybody can harvest without individual punishment. However, other types of property regimes to regulate common resources are related to group, individual and governmental property.

The success of self-organization of effective institutions to control common resources depends on property regimes, as well as evolution of norms and design of rules. Again, this is an excellent example of complex adaptive systems.

Different Ways of Using Models

It should be clear by now that models are not of use to accurately predict future developments of human dimensions of global environmental change. Although models are often used for this purpose, there are more suitable purposes for which they are of importance. In fact, they can be used to overcome the problems raised during the analysis of the current generation of integrated assessment models. Instead of focusing on a single model used for prediction, a combination of different types of models should be used to explore hypotheses and uncertainties. Models can be used in different ways. Here, three different goals of use are distinguished: to understand observed stylized facts, to improve decision-making of complex problems due to interactive use of models, and to explore possible futures by scenario developments.

Understanding Stylized Facts

In the descriptions of the various sectors a number of stylized facts were identified, such as the demographic and health transition, diffusion of technology, and environmental Kuznets curves. These macro-level observations are the result of behavior of agents at smaller scales. To understand stylized facts, and to explore possible changes in macro-level relations, we need to study the behavior of the underlying components.

This type of question relates to current work in evolutionary economics. Various studies analyze what are the important characteristics of firms and technology development to explain structures emerging in specific markets. Examples are the size distribution of firms, and the large number of firms during the beginning of a new market. Such analyses are also frequently performed by various other disciplines, and are assumed to be valuable to understand the underlying dynamics of the observed stylized facts.

Interactive Use of Models

Models are not prediction machines, and should not be used in this way. Moreover, for policy analysis, models should be made available in such a way that they can be used in an interactive fashion. The recent developments in graphical user interfaces, PC availability and Internet, should make it possible to use simple transparent models by a large community of stakeholders. Models can be used in this way as learning tools. Using models in this way is exactly what we can learn from the work of Do¨rner. Dietrich Do¨rner and his colleagues study decision-making in complex situations. Their research design is to develop micro-worlds, which are computer simulation models of a management problem, and to ask real persons to manage the system. These management problems vary from a simple climate-control system to regulation of a virtual society. Because the participants in the game have an incomplete picture of reality, their attempts to manage the system can lead to catastrophes. Do¨rner concludes that managing complex situations requires experience, and this experience can be built up by playing management games, like training pilots in flight-simulators.

Scenario Development and Alternative World Views

Traditionally, scenario analysis starts with an initial forecast which is called the base case or reference scenario. Then alternative assumptions are made of initial conditions, equations and parameter values. The resulting projections are called *scenarios*. Differences between the scenarios and the references are used to evaluate uncertainties and possibilities of policy to influence the future. However, due to the implicit assumptions within every model, one reference scenario will not be enough to assess policy options. In fact, a set of reference cases should be used, which capture the main variety in alternative assumptions.

Scenario analysis should be seen as "computer-aided" storytelling, where different stories can serve as a starting point for analysis. Analyzing policy options for different types of futures, can give insights into the robustness of these policies. Explicitly using different starting points can reduce the illusion that model-based analyses are objective predictions. In such a way, models can serve as a medium for discussion instead of an electronic oracle.

Challenges for The Future

Important weaknesses of current modeling activities, related to human dimensions of global environmental change, are the inherent subjective assumptions of parameter values and relationships, and the fact that macro-relations are used of phenomena which emerge from local interactions between agents.

The challenge in the coming decades will be to use new developments in modeling tools and the use of models to overcome current problems. A promising start can be made when models are used in a more explorative way. Current models are often used as truth-machines. But the predictions, the glimpses of the future derived from electronic oracles, should not be our main interest. Since the future is inherently unpredictable, models should be used to enrich our insights into the behavior of complex reality. Improvements of our mental models can help us to improve our decision-making.

The concept of complex adaptive systems can be a promising starting point to think about new ways to develop and use models concerning human dimensions. Systems evolve, and therefore models assessing the future should focus on evolution and change. This requires the inclusion of disciplines nowadays not highly involved in modeling, such as psychology, institutional science, history and anthropology. Modeling human dimensions of global environmental change is a way of managing uncertainty and

References

Ayres, R U and Simonis, U E, eds (1994) *Industrial Metabolism – Restructuring for Sustainable Development*, United Nations University Press, Tokyo, Japan.

Hughes, B B (1999) *International Futures: Choices in the Face of Uncertainty*, Westview, Boulder, CO.

Janssen, M A (1998) *Modelling Global Change – The Art of Integrated Assessment Modelling*, Edward Elgar, Cheltenham.

Rayner, S and Malone, E L, eds (1998) *Human Choice and Climate Change*, Battelle Press, Columbus, OH.

Rotmans, J and de Vries, B, eds (1997) *Perspectives on Global Change*, Cambridge University Press, Cambridge.

Waldrop, M (1992) *Complexity: The Emerging Science at the Edge of Order and Chaos*, Simon and Schuster, Toronto.

4

Public Attitudes Towards Climate Change

Its particular focus is to ask what we know about existing public attitudes to climate change, whether they are currently evolving, and to provide some pointers to how the impacts of climate change and a move to a 'clean energy future' (e.g. enforcing of low-carbon legislation) will influence public attitudes and behaviour across the world.

Four factors should be noted at the outset, which limit what can be realistically achieved in such a review.

1) The combined academic and grey literature on this topic, while reasonably contained in terms of focus, is considerable. In addition, many of the studies are difficult to compare directly, because of different sampling methodologies, questionnaire wording, and response scales used.
2) Empirical evidence on baseline public understanding and attitudes to climate change in many of the developed nations (UK, USA, Western Europe) is relatively well documented, and forms the basis of a number of the conclusions drawn here. The project scoping document nonetheless asks for predictions of how attitudes will change 'across the world', and this presents immediate difficulties, given that we have only limited comparable survey data for many of the significant developing nations contributing to emissions and/or particularly vulnerable to climate impacts.

3) A consistent theme in any area of comparative attitudes research is that while the basic cognitive processes of individuals can be assumed to be broadly constant across nations and cultures, other important factors such as levels of knowledge/education, values, affective responses and associations, and above all the prevailing social context cannot be assumed to be consistent. Context includes such things as media reporting, social norms, science curricula in schools, local politics, current energy technologies and infrastructure, and is known to be particularly important for shaping attitudes as well as responses to surveys and other types of studies. However, the precise causality of any specific factor is much more difficult to establish.

4) The scoping document asks how might public views evolve in the future? While the long term trend in opinions *to date* is clear, how these will evolve over the next 50, 10 or even 5 years sets a much more challenging task - particularly as the impacts of climate change are themselves uncertain as are the other factors (media reporting and the increased use of the internet for information, world political developments, radical technology transitions) which might also be expected to impact future behaviour and opinions. Accordingly discussion is offered in the final 'Futures' section, of some of the factors which might influence public views in the years to come, although these very much represent personal observations.

Methodological Issues

The report is based primarily upon a review of the existing literature on climate change attitudes and perceptions, alongside examination of evidence from more contemporary studies. Two main sources have been used.

- Published reviews and other material appearing in the academic social and environmental sciences, as well as grey literature.
- This data-base is by no means comprehensive but includes major climate attitude tracking polls in Europe and North America.

Our understanding derives from a range of quantitative studies ranging across traditional opinion polling, experimental studies (primarily within psychology, but also some behavioural economics), and more qualitative exploratory research utilising focus groups, semi- and unstructured interviews (primarily from environmental sociology or geography). Issues investigated by researchers include such things as concern, knowledge, attitudes, trust in scientists, framing impacts and communications, the balance between cognition and affect in reasoning, views of government and governance, temporal and spatial discounting of climate impacts, as well as the relationships with environmentally significant values and behaviours.

It is important to make clear that empirical sources also differ in quality and approach, while diverse disciplinary traditions (e.g. discursive sociology vs. experimental psychology) do not necessarily share comparable methodological or epistemological foundations. What is important, however, is that each study is conducted to meet quality criteria appropriate to its own methodological approach.

- *Quantitative Surveys and Experiments* are better for eliciting broad constructs, and for tracking simple trends over time, but do not always yield identical findings or help us to answer why a participant holds a particular view. Current thinking also suggests that asking people about environmental or technological issues with which they are not generally familiar - of which climate change is a prime example - does not necessarily elicit stable or pre-formed attitudes, but rather that respondents are forced to *construct their preferences* based upon a range of broadly sensible, but also contextually sensitive, factors: such things as pre-existing knowledge and associations, direct observation and analogies, personal values, perceived societal norms, and cues present in the question framing itself. As a result question wording can be a major influence on survey responses, and tracking should only be done where identical questions are used and, ideally, identical sampling and survey administration procedures. For example, there is now evidence that people respond somewhat differently when asked about 'global warming', a concept more in keeping with the initial public discussions of the issue in the 1960s and 70s (and the term still most commonly used in surveys conducted in the USA), as compared to when asked about the more generic idea of 'climate change'.
- *Conventional Qualitative Studies* utilise established social sciences techniques such as semi-structured interviewing or group discussions. Systematic methodologies for eliciting and analysing qualitative data have a very long history in the social sciences and offer, in contrast to surveys, far greater depth and insight into people's own understandings of climate change, the frames and discourses people use to structure

those understandings, and as a result can help to answer the underlying 'why' questions. By the same token it is more difficult to generalise findings from qualitative research, let alone to track changes in discourses over time.

- Finally, *Deliberative Approaches* tend to combine provision of structured information about the topic with group discussions (often involving domain experts), and can also incorporate value elicitation and decision-structuring techniques. Such approaches are particularly useful for developing recommendations, identifying key trade-offs, and formulating policy advice through extended processes of deliberation and dialogue between laypeople and experts. Properly conducted deliberative processes take place over an extended period of time, and allow participants to develop and modify their views as they gain information about the topic. Consequently they are very costly to design, run and analyse, and tend to be rarer than surveys or more conventional qualitative approaches to eliciting views. There are relatively few well documented examples for climate change;

There is considerable value in adopting a mixed-methods approach to climate attitudes research, by combining evidence from quantitative, qualitative and deliberative studies – as this often gives us a more complete picture than any one approach would alone, and is the strategy adopted here.

Concern and Importance

We forget sometimes that climate change is not a particularly new issue for the public. Levels of concern about climate change have been increasing significantly since the early 1960s, particularly in Western European countries, where relatively high levels were reached even as far back as 20 years ago.

In the European Union 76% of respondents in the 12 EC Member States were very / somewhat worried about the greenhouse effect in 1988, increasing to 89% in 1992 (INRA).. Similarly, the 1992 Gallup Health of the Planet (HOP) survey indicated that more than half of the respondents in 13 out of 24 countries worldwide felt that climate change was a serious problem, although of these 13 nations, 8 were European. In the ensuing two decades reported concern about climate change, when asked directly, has become widespread in Europe, the USA and elsewhere. Concern is also particularly high amongst the young, with a 2008 survey of 12-18 year olds by the UN Environment Programme in five countries finding very high levels, and greater optimism that something could be done about climate change, as compared to adults in the same countries.

A further consistent key finding is that where judgements of other issues are also solicited climate change is invariably not the highest or most important priority for many people – other global/societal (world poverty, crime, terrorism, wars etc.), environmental (water pollution) or personal issues (health, finances, relationships) typically have a higher expressed importance for people. Theoretically, people may have only a 'finite pool of worry' to draw upon when considering all of the potential issues that they might attend to in life. This is important because any of these other issues may push environmental/climate concerns temporarily aside as personal circumstance, national events or media reporting make them more or less salient for people. Equally, some of the other issues which people care deeply about (long-term economic prosperity) are in fact related to climate change – but only in indirect ways.

Knowledge

We know that there is now widespread awareness that climate change is happening. People also recognise that humans are in part responsible for this. In a representative British survey conducted early in 2010 Spence, Venables et al find that 31% believe climate change is mostly or entirely due to human activity, with a further 43% believing it is partly due to human activity and partly natural processes. By contrast, people's detailed knowledge is less certain, which they themselves recognise. When discussing 'knowledge' of climate it is important to recognise that climate change itself is a complex phenomenon, with multiple and interdependent causes (both social and environmental), areas of scientific knowledge which are still incomplete or contested, and impacts which are uncertain and often far in the future. Nor is climate change observable to individuals in any direct sense: even experts only understand it second-hand, through the concepts and language of science. As a result it is particularly difficult for people to conceptualise and relate climate change to their daily activities because as a complex science construct it is not easily translated into the language of popular culture. Laypeople ordinarily form attitudes and beliefs through their direct experience and familiarity with a person, object or phenomenon, or through information received from mediated sources (family, friends, TV, print, and latterly the internet). Not surprisingly, then, many studies have shown that people tend to associate climate change with their most immediate of direct observations: that is, of daily or seasonal fluctuations in weather, something at odds with the scientific concept of 'average weather conditions (temperature, precipitation, atmospheric etc.) as measured over an extended period of time'. Conflation of 'climate' with 'weather' contributes in no small part to the belief that climate change has natural rather than human causes. Other systematic misconceptions have also been documented.

Early qualitative research conducted in the USA already gave some indication of how people's mental conceptualisations of climate change diverge from scientific understandings. Kempton's study illustrated how US citizens interpreted climate change in terms of four pre-existent categories related to: stratospheric ozone depletion, plant photosynthesis and respiration, generic 'air pollution', and experienced temperature/weather variations. Bostrom et al likewise found people's mental models confused climate change with ozone depletion. Both issues rose to prominence at about the same time in the 1960s, and it is easy to see why people might reason that the hole in the ozone layer is letting through more sunlight and heat when prompted to describe a causal mechanism for climate change.

Generally, individuals in the past have also held a limited understanding of the human contributions to a changing climate. Brechin's analyses of surveys undertaken in 1999 and 2001 show misunderstandings world-wide, even in nations considered to have strong environmental values (e.g. Germany). Of the respondents in 27 countries surveyed in 1999, very few indicated that burning of fossil fuels was the main anthropogenic contribution to global warming. The survey also shows that many respondents indicated deforestation and 'air pollution' as causes. Similarly, in the British study by Poortinga, Pidgeon and Lorenzoni air pollution was the most common cause of climate change cited by people unprompted. A mental model of generic 'pollution' would not necessarily help people to discriminate between the actual causes of climate change (burning of fossil fuels) and incorrect ones (e.g. particulate and sulphate emissions from cars and factories, which, while undesirable for other reasons, both serve to *reflect* sunlight). A qualitative evaluation study by EcoAmerica concludes, however, that discourses around 'preventing pollution' and 'protecting the air that we breath' do form persuasive arguments for many people, particularly when they are contextualised against a backdrop of improving national energy security. Such framings essentially work with (rather than against) people's pre-existing mental models of climate change.

Another common misconception reported for some time in the perceptions literature is the association with generation of energy from nuclear power stations. Upham et al also point out that alongside misconceptions there are some gaps in people's understanding of climate change causes, since few respondents in surveys mention domestic energy use, meat eating or food transportation as causes. Brechin concludes that knowledge across the globe has improved since 2001 – particularly with respect to the role played by fossil fuels. This almost certainly reflects the greater attention now being paid to climate change issues in the global media, school curricula and civil society more generally in many countries.

Finally, knowledge of the consequences of climate change has tended to focus on its immediate physical impacts rather than other forms of social disruption, human health impacts, or secondary environmental outcomes such as rising ocean acidification. When asked about impacts people also tend to draw upon the associations they make with weather. Upham et al note that a debate now exists in the literature as to whether linking climate change to extreme weather events prompts people to greater concern and personal action regarding climate change (as something significant to be responded to) or alternatively to dismiss the seriousness of the issue, with the evidence mixed on this point.

Not all misconceptions will matter of course. Rather, experience from risk communication research suggests that policy makers should concentrate upon reinforcing correct beliefs and challenging incorrect beliefs that might directly influence relevant behaviour (e.g. if someone believes that recycling domestic waste should be their main response to climate change), while at the same time stating the most effective actions people can take.

In methodological terms, there are longstanding arguments over the validity of direct attempts to measure lay-people's knowledge of a science or environmental risk topic through structured questionnaires. The correct phrasing and calibration of questions is one challenge (too difficult and everybody looks stupid - too easy and everybody looks smart!), as is the interpretation of what constitutes a 'correct' responses where there remains legitimate scientific disagreement over a topic. Some have even argued that the very selection of traditional science questions displays a cultural bias, in privileging scientific knowledge as the one normatively appropriate 'truth'. For example, one could be very knowledgeable about aspects of climate change policy or economics, without necessarily knowing details of the underlying physics or chemistry.

In a related set of arguments, scholars within the tradition of sociology of scientific knowledge have also mounted a robust critique of the 'deficit model' of science communication: the simplistic idea that opposition to new developments in science and technology (nuclear power, biotechnology, stem cell technology etc.) arises merely because people lack the appropriate scientific training or knowledge. The simple conclusion which would follow, were this true, is that a greater effort should be made to educate people in science. While increasing scientific literacy is, of course, important for many reasons, researchers have pointed out that concerns about risks and the environment are not necessarily a function of ignorance, but often reflect differing cultural and value positions, or assumptions about wider institutional arrangements such as distrust in the governance and management of technology and its risks.

The principal conclusion to be drawn from these extensive debates about knowledge is that, whilst climate and carbon literacy may be desirable things in and of themselves in a modern society, merely making people more concerned about climate impacts, or more knowledgeable about the science, is unlikely to lead to changed behaviour or different attitudes. Indeed research has shown that there are a range of powerful influences, over and above lack of knowledge, which act as barriers to public engagement with climate change. These include: individual factors such as cognitive dissonance reduction (between espoused beliefs and current actions), perceived uncertainty, habits, feelings of fatalism and helplessness; and social factors such as lack of enabling initiatives, prevailing social norms, and lack of visible political or government action. One way of addressing these barriers is clearly though individual or community engagement and behaviour change initiatives at a local level, although these will have to be combined with appropriate structural, legal and fiscal instruments. To this can be added the objective of fostering collective action across more extended spatial scales utilising the interconnected networks that now extend across our globalised world. Another route to change is to exploit that we know about existing actor-networks in the climate policy domain to link individual perceptions with actions at the wider institutional level. As

Ockwell, O'Neill and Whitmarsh argue, one aim of climate communication and attitudes change efforts should be to foster grass-root demands for *institutional action*, in this way fostering direct public support (and political space) for more radical policy and government interventions.

Perceptions of Risk

Three decades of work on the psychology of risk perception indicate that the balance of perceived risks with tangible benefits is one of the factors which drives overall risk perceptions and judgements of a risk issue. At a very general level, asking about people's concern is also a proxy measure for perceived risk. When asked directly, most people feel that climate change holds risks and that these outweigh any benefits. In a survey in Britain in 2002 fully 55% thought the risks of climate change outweighed benefits, compared to only 14% who thought benefits outweighed risks. However, respondents also differentiated between aspects of the risk, characterising climate change as a moral issue, one with risks for future generations and something they did not have personal control over – all factors which have been associated with heightened levels of risk perception. As noted above also, the many facets of climate change are complex, making any evaluation of 'risk' particularly difficult even for expert analysis, while the level of acceptable risk is in part a matter of the values particular groups in society hold - hence different individuals might legitimately disagree about what is acceptable.

Opinion polls and more in-depth studies also show that individuals distinguish between the effects of climate change on their personal lives and on wider society – with judgements that risks are higher for people who are spatially or temporally distant. Potential hazards arising from climate change to society are evaluated more highly than individual threats while several studies show that, although most Europeans are aware of the potential risks of climate change world-wide and the adverse consequences that may befall societies in general, they tend to attenuate the risks to themselves personally. Optimism biases such as this are not unique to climate change of course, being found with a range of health and other risks. In the USA, Bord *et al.* also conclude that global warming is generally not perceived to be personally threatening. Among a sample of US citizens, who were asked to rate the likely threat to themselves personally (i.e. during their lifetime) of various social and environmental issues, global warming featured lowest in perceived personal threat, while heart disease, cancer and car accidents were listed at the top. The authors therefore argue that respondents distinguish the personal and societal implications of threats, and that most of the non context-specific surveys which indicate high concern about climate change are in fact eliciting concern about societal impacts.

How risks and benefits are framed is also important. Some studies have shown that if benefits associated with current lifestyles are considered these can be felt to outweigh, on a personal level, the possible risks of climate change. When the overall effects on society are considered, these benefits are still considerable, but the perceived degree of harm is also higher.

Perhaps the most important result from the risk perception literature is that climate change is a temporally and spatially distant issue for many people. That is, it is seen to threaten people in other places of the globe or those born far in the future. As a result people do not always view climate as personally threatening or relevant to them, with one consequence that this may prevent people from forming strong affective associations with the issue. To some extent such judgements are a true reflection of the very real commons dilemma that climate change poses us all – something caused by small individual actions (which are personally and immediately beneficial) with wide ranging but diffuse and uncertain future global consequences. A number of psychological explanations can be advanced for this; evidence of any personal threat from climate change is difficult to come by in our day to day lives, or we might choose to deny that our actions will cause us future personal losses. Equally, psychological theory suggests that people construe future events in more general terms while immediate losses are construed in more concrete terms. Finally, as Swim et al point out, we know from a long history of experiments in behavioural decision making that people at times place steep discount rates on costs born in the future compared to equivalent costs that will be born today. Such discounting effects are, however, also known to be context and task dependent. Whatever the root drivers of this phenomenon, temporal and spatial distance poses a serious problem when attempting to motivate people to act now on mitigation.

Responsibility for Action

In nationwide surveys conducted in Britain in 2005 and 2010 only 8% and 10% respectively of respondents stated that the main responsibility for action on climate change lay with individuals and their families. Rather, primary responsibility was overwhelmingly ascribed to powerful external actors such as the international community, national governments and businesses. In part this can be interpreted as people seeking to displace responsibility for major action onto others rather than to themselves as individuals, and in this way avoid what are perceived to be costly or difficult changes to behaviour and lifestyles. Qualitative research also confirms that people see responsibility for tackling climate change to lie primarily at the state actor level, even though, almost in the same breath, people will also state that they do not fully trust those actors. For many the issue is seen as too big a problem for them as individuals to even begin to tackle themselves, with inter-governmental agreement also seen as necessary to avoid other countries free-riding. Equally, politicians who may fear the electoral cycle are restrained from taking tough action on climate change believing they will be punished at the ballot box if they attempt to force people to accept radical change in lifestyles, new environmental taxes, or stricter legislation. This represents a complex and serious 'governance trap', with governments and governed seeking to lay responsibility with each other, and is at odds with current UK government policy which, alongside the adoption of low carbon technologies, also aims to foster individual responsibility and behaviour change in pursuit of energy demand reduction.

When asked whether they are personally prepared to do *something* about climate change the majority of people respond in the affirmative, although many will state (a) that it is currently difficult to take appropriate action but (b) that they would respond if only given the right guidance, help or incentives from government. There remains, then, a clear gap between peoples' expressed concerns and their willingness to change their behaviour significantly (the so called value-action gap). Also what people say they propose to do to help with climate change (e.g. increase their domestic recycling) might not always be the most efficient for cutting personal carbon emissions. Research on this problem from other domains such as health psychology, reviewed in Spence and Pidgeon, suggests that habit is an important determinant of current behaviour, and that long-term behaviour change requires both effective 'downstream' communications to highlight why changing behaviours is important, coupled with 'upstream' structural changes (infrastructure, legislation, or financial) to encourage or force a different set of practices and patterns of behaviour. Changed attitudes then tend to follow the changes in behaviours, with examples of successful coupled interventions including the campaigns to increase seat-belt use in vehicles, or to restrict smoking in public places. However, in both of these cases it took many years to identify the problem, devise effective communication and structural/legislative changes, and fully change behaviours across society – a luxury we may not now have with climate change.

Values and Culture

Some researchers have segmented populations empirically along dimensions of concern and willingness to act, by using statistical clustering techniques to identify sub-groups of society: from those who are fully engaged with climate issues through to those who are doubting or denying. Leiserowitz and colleagues segment US respondents into six distinctive groupings, ranging from a small and highly engaged 'Alarmed' group at one extreme (10% of the population in early 2010) to a 'Dismissive group at the other (16%). In between are 4 groupings who all express some form of ambivalence about climate change ('Concerned', 'Cautious', 'Disengaged' or Doubtful: totalling almost 75% of the sample) with the implication that it is these individuals who might be most open to change in both their beliefs and behaviour. In a study conducted simultaneously in Italy and the UK Lorenzoni and Hulme report a similar spectrum of beliefs covering Denying, Doubting, Disinterested and Engaged. However, a disadvantage of statistical segmentation approaches is that they result in categories which are hypothetical (and hence should always be treated in that way), rather than in social groups as we customarily understand them. It is also difficult to relate such groupings to more theoretically validated cultural or ideological factors.

Regarding wider cultural factors, there is evidence that attitudes towards climate change are influenced by people's broader environmental beliefs, as well political and ideological values, although the precise relationships are not always clear-cut. Generic concern about the environment, as measured for example through standardised instruments such as the New Environmental Paradigm (NEP), is positively related to concerns for climate change and greater support for climate measures. Such a relationship is to be expected, given that climate change is, for many people, symbolic of the wider threats posed to the environment by human activities and industrialised society, and would be predicted from the values-beliefs-norms (VBN) model of environmental behaviours.

In terms of traditional partisan politics, we also know that supporters of right-leaning parties tend to be more sceptical of global warming claims than supporters of left or green-leaning parties. In the USA, in particular, there is now a very wide gap between the beliefs of Republican and Democrat voters. This difference is not simply a reflection of differing general philosophies toward the role of government. Partisan differences were negligible before the autumn of 1997 when the 2nd Clinton administration launched a campaign to build public support for the Kyoto Treaty.

After the barrage of media coverage at those time, partisan differences increased significantly according to Krosnick et al. Dunlap and McCright argue that the roots of this difference in US public views lie also in the actions of libertarian and right-leaning commentators, as well as the activities of the petroleum industry and affiliated parties, who have sought over the years to systematically highlight uncertainties in climate science. Dietz et al also point out that political affiliation, while a strong predictor of policy support for climate measures in the USA, exerts its influence only indirectly. In particular, left-leaning individuals have higher environmental values and greater trust in statements and policies proposed by environmentalists, which are in turn both positively related to their support for climate measures.

Some evidence also exists that the cultural discourses of 'individualism' and 'egalitarianism' as described by the cultural theory of risk, which are known to link particular ways of life with perceptions of environmental risk, can also operate in relation to climate change. The theory argues that individualists fear threats to market choice while egalitarians fear threats to shared resources and communitarian values, including those associated with the environment. Cultural position is also known to influence the interpretation of new ambiguous information about climate change and other risk issues, in ways which serve to confirm individuals' pre-existing beliefs, hence fostering resistance to attitude change or even leading to a greater polarisation of views for those already at the extremes of opinion.

The Social Amplification of Risk

As with many other environmental issues, the changing social context at any particular point in time (e.g. the activities of particular interest groups, or media reporting, or just the onset of events such as an exceptionally cold winter or hot summer), can serve to amplify or attenuate perceptions of risk. Various case studies have shown that events related to hazards interact in highly dynamic ways with psychological, social, cultural and institutional factors resulting in amplification or attenuation of individual and social representations of risk and danger. A key insight from that literature is that amplification or attenuation is rarely driven by any single factor – rather a series of factors need to be present in combination. Until quite recently climate change has, in lay perceptions, been an attenuated or 'hidden hazard'. In the Kaspersons' words, climate change became attenuated because it was both a value-threatening and an ideological hazard. In the US, and to some extent the UK, the actions of commentators and companies to undermine the case for climate change have also contributed to this situation.

However, hidden hazards may not remain so forever, and a key element of social attenuation and amplification processes is their dynamic nature, coupled with the way in which different institutions, such as the mass media, constantly seek to transform both scientific information and policy debates about risks. For example, Carvalho and Burgess analyse reporting of three UK broadsheets over 18 years and explain how communication about climate change is a multi-directional process delineated and shaped by the social learning of the producers and receivers of the information. The authors identify two main influences on the newspapers' discourses: the ideological stance of the reporting media themselves, and the political positions encapsulated by senior government officials' pronouncements on this topic. Smith also points to the tension between the practices and norms inherent in professional news journalism and the characteristics of trans-disciplinary issues such as climate change, which require long-term, consistent and cross-cutting coverage rather than reporting in the blazing and eye-catching form of a news-story, often by dramatising scientific uncertainty and political controversy, something we discuss briefly below. The distinctive framings being brought to bear on climate change within the media have also begun to receive significant attention from researchers.

It is no longer clear whether climate change can still be classed a hidden hazard. One would expect this situation to be changing given the amount of UK media and policy coverage that climate change has received since 2006, although against this many of the ideological and value-threatening aspects of the issue remain unchanged and most likely still act as a brake on increases in concerns and risk perceptions associated with this issue. As the events of the winter of 2009-2010 have also shown us, external issues such as the inconclusive Copenhagen talks, the East Anglia e-mails controversy, and reported errors in glacier melting forecasts in the last IPCC report, can all arise in unpredictable fashion to impact public perceptions. Investigations of the dynamic nature of climate risk perceptions, and the factors which might be causing them to change or not, is clearly an ongoing research need.

International Comparisons

There is a small but expanding corpus of research directly comparing attitudes to climate change across different nations. The methodology behind such comparative work is difficult, costly and it remains the case that such surveys, almost by definition, contain a very restricted set of questions - typically awareness and knowledge items, measures of concern, and evaluations of general policy options. The comprehensive reviews in particular by Brechin and Leiserowitz ground a number of the main conclusions reported here, and readers are referred to these two sources for more details of the precise polling cited. To obtain an overall picture the results of several of these polls are considered in some detail.

The cross-national evidence is particularly fragmentary, with no one in-depth study being replicated in its entirety across more than one nation. The problems with surveys are writ large in cross-national efforts, in particular being reliant upon a restricted set of questions, and lack of comparability from poll to poll. Notwithstanding this, the evidence is that levels of knowledge and concern have both increased globally over the past 15 years, although climate change remains secondary in importance for people (and this is especially so since the onset of the global financial crisis in 2008). Possibly more significant is the finding that key country differences on some aspects have emerged over the past 15 or so years, in ways not simply attributable to national levels of education or development. It is useful therefore to take a temporal perspective on the findings.

Data from the earliest cross-national poll, the 1992 Health of the Planet Survey, show a relatively limited awareness of climate issues, with other environmental issues such as air and water pollution being of greater concern. Respondents in this and other national studies of the time also showed very limited understanding of the causes of global warming, especially the role played by fossil fuels. Likewise, in the 1999 GlobeScan survey of 25 nations depletion of ozone was the single most cited cause of the greenhouse effect, followed by air pollution from factories and cars. In a very few nations (e.g. India, Turkey, Argentina, Malaysia) loss of forests was cited as the main cause.

Brechin reports that in the Environics International study of 15 nations conducted in 2001 understanding, while still low, had improved somewhat but with no large differences in climate change knowledge between developed and developing nations. Support for the Kyoto protocol was, however, higher in European countries compared to in the USA, in part reflecting the emerging transatlantic differences in the politics of climate change. Concern was also much lower in the US than in many other countries around the globe. As discussed above, political polarization had begun to emerge in public attitudes within the USA just prior to this (Krosnick et al, 2000; Dunlap and McCright, 2008). A GlobeSpan survey in 2001 found evidence that, with the sole exception of China, people in developing countries were far more worried about climate change posing a 'direct threat to you or your family over the next decade' compared to either European or US respondents. As Leserowitz comments, this latter finding most likely reflects (a) the reality of long-term vulnerability in adaptive capacity in many developing nations, as well as (b) the belief held by people living in Europe and the USA that the most severe impacts of climate change will fall on other people and at other times.

Some detectable differences between European nations do appear to have begun to emerge at the turn of the millennium. Lorenzoni and Pidgeon report greater concern about climate change amongst the Southern European nations in Eurobarometer polling conducted in 2002 – something which probably reflects the associations respondents made between climate change and localised weather impacts, and a realistic assessment of the future impacts on weather of increasing temperatures in the Mediterranean region, compared to, for example, those envisioned for Scandinavian countries or northern Scotland. This study therefore suggests that we should not overlook the importance of geographical and cultural influences, even within a single nation or region, on public perceptions of climate change.

Between 2000 and 2006 those reporting climate change a 'very serious threat' had increased substantially in research by GlobeScan, with majorities now in 17 of the 22 countries polled. A similar trend is observed from the

Pew Global Attitudes Project which shows much higher concern for the 'environment in general' in 2007 compared to 2002. Regarding whether climate change is a very serious problem, respondents in some of the most vulnerable nations (e.g. Brazil, Bangladesh) voiced the highest concern in the 2007 Pew poll. However, when compared to many of the other nations polled, far fewer people in Britain, the USA, Russia and China considered climate change to be a very serious problem. Brechin points out the worrying implication of this – that the latter three are the World's biggest producers of greenhouse gasses. The finding of rising concern (and perceived seriousness) over the last decade also has to be set against the fact that climate change remained of secondary importance to economic and other problems in most nations. Brechin also concludes, more optimistically, that the period 2001-2008 saw improved understanding of the issues across the globe: in particular regarding the role played by burning fossil fuels, and a clear decrease in belief in ozone as a cause in several surveys. However, this improvement in knowledge is variable, with much lower levels of self-reported knowledge and awareness remaining in many of the poore nations of the world (also found in a 2009 Gallup survey reported in Pugliese and Ray, 2009), accompanied by lower levels of support for climate action and policies when knowledge is low.

Our understanding of attitudes and beliefs in many of the developing nations most vulnerable to climate impacts is particularly patchy. Some contextually-grounded qualitative work does exist from a development studies perspective, but tends to focus upon climate change as one of a range of environmental stressors, as well as local adaptive capacity and response. A survey of 132 residents in Nairobi, Kenya concluded that climate change appears as a 'drop in the ocean' compared to the problems of poverty, unemployment, crime and corruption. The qualitative study *Africa Talks Climate* in 10 African nations reports that, while most participants were aware that their weather was changing, there were low levels of awareness and knowledge of the causes and impacts of climate change, especially amongst rural dwellers. However, drought, deforestation and flooding were key environmental concerns for many Africans, while respondents in rural areas felt that migration would be their only practical long-term response to climate change. Common frames of reference used for understanding the issue were the emphasis placed upon trees (which many Africans believe will attract rain), faith or religion (participants in some countries felt that the will of God determines adverse weather), conflation with the ozone problem and general 'air pollution', and the direct heating effects of local activities. Additionally, Brechin reports that in India, where climate vulnerability and public concern are both high, support for climate policies is relatively low, while both China and Russia have low levels of public concern comparable only to the USA. All of these studies illustrate that, while we might now know enough to explain patterns of beliefs and perceptions currently in the USA (and much of Western Europe), many localised contextual factors, over and above education and development levels, are needed to explain the considerable variation in knowledge and support for policies in many of the significant nations of the world. Accordingly, both Brechin and Leiserowitz recommend further detailed cross-cultural research on the factors - psychological, cultural, geographical, economic and political -which might help us to explain differences in patterns of attitudes and knowledge in significant regions of the world.

A number of the international surveys also ask about support for a variety of climate measures – although again the types of measures and question wording differ in different polls. Summarising these studies, respondents across a range of nations often express a strong desire for action now on climate change (as opposed to a 'wait and see' approach) on the part of national governments and the international community. Support for the Kyoto protocol, and for the development of low carbon and energy efficient technologies, has also been high in many nations. Brechin documents results from the 2008 WorldPublicOpinion survey of energy preferences, where a clear majority in 21 nations favoured deployment of renewable energy (solar and wind) – a result consistent with recent detailed surveys of energy preferences in Britain. Attitudes to nuclear energy by contrast differed markedly across nations, ranging from 14% of respondents in Germany wanting more emphasis on nuclear (and 63% wanting less), to China where fully 63% wanted more emphasis (and only 10% less). In Britain people appear more evenly divided on the question of new nuclear power (41% more emphasis versus 29% less), a result again consistent with detailed national survey and site-based work. The WorldPublicOpinion survey also frames several of its questions in terms of the costs of policies. Although mention of costs does reduce the attractiveness of some options, levels of support for renewable energy remain robust even when people are informed of initial higher costs (many believing that renewable energy will be cheaper in the long-run). Commenting on this, Brechin concludes that there may be more of a consensus amongst the public cross-nationally on the need, and means, to face up to difficult climate and energy choices than is often assumed by policymakers. A 2007 BBC World Service poll of 21 countries finds that large majorities in all countries think that people will definitely or probably need to change their lifestyles to reduce emissions. It also asks about raising taxes to curb the use of fossil fuels – and finds, not surprisingly, mixed results. What is significant in this poll, however, is that support for such taxes, averaged over the 21 nations, rises substantially (from 50% to over 75%) if their use is specified: either to improve energy efficiency, or to lower other taxes so that the overall burden remains the same. Leiserowitz similarly reports that a 2001 GlobeScan survey found strong support across 28 nations for a rise in petrol prices 'if the money were used to cut air pollution', and concludes that under the right circumstances many publics would be prepared to consider higher fuel prices. This points out, once more, the ways in which different framings for tradeoffs in surveys can elicit different responses, and also suggests a need for further research, specifically into the conditions under which particular monetary trade-offs to tackle climate change might be acceptable (or not) to people in Britain and elsewhere.

More in-depth deliberative and qualitative work allows participants to form a considered opinion when presented with information about an issue. However, we have very few comparative studies using such methodologies, not least because of the difficulties of conducting such work simultaneously in two or more nations and then developing suitable and systematic comparative data analysis. One exception was the cross-European ULYSSES integrated participatory assessment project – which sought views on climate change in several European countries using focus group and other techniques during 1997-8. This research showed that participants wanted more information about climate change, but that they also thought that governments should go further than the provisions allowed in the Kyoto protocol.

A major deliberative exercise coordinated by the Danish Board of Technology – unique in scale and number of countries involved – ran simultaneously in 38 countries across the world on September 26th 2009, in the run up to the COP15 Copenhagen conference. Groups of 100 people met in each country, and were provided with the same information, comprising a detailed 40 page booklet and a 10-15 minute video. They then debated possible policy responses and voted on a range of predefined questions (covering climate consequences, urgency and long-term goals, dealing with emissions, and the economy of technology and adaptation) as well as questions that participants themselves defined. As such this can be considered a large-scale experiment in climate risk communication and information provision. The organisers report that what stands out clearly across the exercise is that these citizens, as a result of learning more about climate change and debating the issues involved, wanted strong and rapid action from their governments at the COP15 negotiations, recommending also that their governments follow this up with reductions beyond current commitments. They also called for enabling initiatives such as information provision and positive market incentives to motivate citizens to act, international action and commitments, fair burden sharing between poorer and richer nations, and an international financial system to pay for technology transfer. Participants were also prepared to accept rises in the price of fossil fuels, although many felt that the burden for this should fall on richer countries.

The findings from both of the cross-national deliberative studies outlined here – while occurring more than 10 years apart – are congruent with the findings on policy support from the cross-national polling. Taken together they indicate a significant underlying public desire for decisive action from both national governments and the international community. They also point to the usefulness of employing more deliberative approaches to eliciting citizen views on global climate policy.

Changing Attitudes

2007 saw the publication of the IPCC 4th Assessment Report which stated with a high degree of certainty that anthropogenic activities were altering the earth's climate, as did the prominent 2006 UK Stern Report on the economics of climate change. While one might expect these reports and surrounding discussions to be followed by a further *increase* in concern, perversely this seems not to have been the case. If anything scepticism, and with it lowered concern about the anthropogenic causes of climate change, has been gradually rising between 2007 and mid-2009. In the UK in August 2009 only 76% were very or fairly concerned about climate change compared to 81% in August 2006 on an identical tracking survey for the

Department for Transport while similar trends have been reported in Europe, the USA, and in cross-national polling. Note that all of these studies, showing a gradual decline in concern, were conducted *prior* to the onset of the climate e-mail controversy of the winter 2009-10, a point which needs to be born in mind when interpreting results from more recent studies.

Potential reasons for this gradual trend since 2007 include:

— *Issue fatigue*, as climate change has come to be extensively discussed in the UK and worldwide media following the IPCC 4th Assessment and the Stern reports. At the same time it is increasingly used as a justification for policy proposals in a wide range of sustainability domains not all immediately recognisable as being climate-related (transport, housing, food, social, defence and overseas aid policy etc.).

— *The impacts of the 2008-9 financial crisis* on the personal importance of climate change and the environment. This might be because environmental issues are temporarily driven out of sight for many people by more pressing concerns (the 'finite pool of worry' hypothesis. Alternatively people may believe, rightly or wrongly, that economic

— *Cognitive dissonance*, expressed as a resistance to home truths as the anthropogenic nature and scale of climate change impacts becomes much clearer. A collective failure to admit that lifestyles today might now need to change in radical ways may well have combined with a degree of denial brought about by some of the more catastrophic and emotive portrayals of future climate impacts.

— *Politicisation and distrust* of communications about climate change - with less trusted actors (politicians, media, major companies) now far more prominent than scientists in communicating with people about the issues, particularly as the debate moves on from the underlying science to consider policy responses.

— *'Climate sceptic' agendas* gaining greater prominence, as various groups and individuals continue with their long-standing attempts to highlight uncertainties in climate science. For Britain this hypothesis is less plausible for the period prior to the onset of the University of East Anglia e-mail controversy in late November 2009, but perfectly possible as an explanation for further changes in beliefs after that point. Prior to the East Anglia controversy, reporting of climate change in the mainstream UK media following the Stern and IPCC 4th reports had more or less

— *Uncertainty transfer* - a very subtle process, where uncertainty over one aspect of climate science and policy spreads to generate uncertainty and scepticism about a different aspect. In particular, the increasing discussion of the many uncertainties associated with climate impacts and adaptation measures may, perversely, increase the uncertainty that some people feel over the reality and anthropogenic causes of climate change. Although not necessarily a serious issue at this point in time, such uncertainties are increasingly being framed explicitly in risk terms for policy and decision-making purposes, as for example with the UK Climate Projections '09. Hence, this may well set a number of very difficult challenges for climate communicators over the coming decades.

In addition to the medium-term rise in scepticism, the mass media are known to be a source of attenuation and amplification of risk perceptions. The very concerted attention during the debate about the University of East Anglia e-mails and the IPCC glacier assessments in the winter of 2009-2010 would also be expected to further impact beliefs about climate change, by adding to the recent decline in belief. A BBC/Times Populus poll conducted in December 2009 and then repeated in February 2010 appears to confirm this, finding a clear short-term decline in the numbers in Britain who believe that climate change is happening, while a YouGOV/EDF poll conducted in May 2010 also finds a drop of one quarter in those who think climate change is a serious and urgent problem needing radical steps. Other more comprehensive surveys conducted in early 2010 in both Britain and the US also seem to confirm a decline in belief, with less people now believing in climate change and its seriousness, across a range of measures, than did a few years ago. It is important to note here that, in interpreting the latter two studies, the longer-term influences are likely compounded with any impacts of the short-term media reporting. Additionally, the UK survey which was conducted by Ipsos-Mori for Cardiff University between January and March 2010, while the e-mail controversy was at its height and in the wake of the unusually harsh UK winter, found that fully 78% still believed that the world's climate was changing. The climate attitudes hour-glass might be not now be half-empty as some have attempted to claim, but despite everything remains over three-quarters full, a point echoed about US public opinion recently by the political scientist Krosnick. What we do know about events such as the climate e-mails affair is that the impacts of media amplified stories and controversies such as this are often relatively short-lived in terms of impacts on attitudes and behaviour (e.g. the change in travel patterns following both the September 11th and the July 7th terrorist attacks were transitory) and confound any longer-term trends which might be present. This suggests that, without further major controversy, we might expect belief in climate change to gradually recover to the earlier trend levels - something already beginning to happen in US data collected in early June 2010. We shall not know the full impacts of these events on attitudes for some time yet, and whether any will prove permanent. Amplification events are, however, also known to have secondary consequences, and one unfortunate outcome might well be for some UK media to revert to its traditional practice of 'balance' in reporting climate stories – hence, for a period into the future, giving a false impression regarding the overwhelming scientific consensus on the anthropogenic causes of warming. It is also the case that the main effect of many of the media reports about the e-mail controversy might simply be to reinforce the existing views of those in society who have already take up sceptical positions, rather than to prompt any fundamental shift in the attitudes of the far larger segments of ambivalent and concerned groups amongst the British or global publics – many of whom have probably not even heard of this controversy at all.. Again, further research is warranted on this important issue.

Governance, Climate and Sustainability

While there remain gaps in the evidence-base, for example with respect to developing nations, the pattern is a fairly consistent one. However, how attitudes will evolve over 5, 10 or 50 years sets a much more difficult task - particularly as the impacts of climate change are themselves uncertain, as are the important contextual factors such as media reporting, political events, decarbonisation and adaptation efforts, or future development of geoengineering 'solutions', all of which might impact opinions.

A first point to make is that a considerable challenge already exists for both governments and communicators in making climate change a salient issue for people in the UK – one which is sufficiently motivating to make mitigation a focus of everybody's lives and actions. There currently exists a very real gap between people's expressed concerns about the environment and their actions (the value-action gap). What also seems clear is that context matters in constraining action and belief. Context shapes and constrains both attitudes and behaviour, and what is possible both now and in the future. If the relevant context does not change (or changes only very slowly) then attitudes and behaviour towards the environment are unlikely to undergo radical change either.

Will changing climate impacts matter in shifting perceptions, as the first and most obvious contextual variable? In the short-term probably not, for a number of interrelated reasons. Currently, as discussed above, people view climate impacts as distant from them personally in both time and space. They also often associate 'climate' with their own direct observations of the weather. A plausible hypothesis here is that, as disrupted UK weather patterns become more evident through media reports and direct observation, particularly if a series of such events occurs relatively rapidly, people will begin to see climate impacts as personal to them. Estimates differ on the extent of likely global mean temperature rise over the course of the coming century, although it seems increasingly likely that without radical mitigation measures global temperature change will not be limited to the 2° rise conventionally seen as the threshold for 'dangerous' climate change. We may well see rises of this order as soon as 2050, and further mean global temperature rises that breach this limit (say 3-4° or worse) by 2100. Even under such a scenario, observable physical impacts in the UK during the first half of this century would involve a gradual rise in the frequency of prolonged hot dry spells in summer, more instances of water restrictions, increased (but localised) flooding events, coupled with a greater frequency and intensity of rain in the West and North West. Some aspects of land use might also have to change over time. Many of these impacts will be difficult and expensive to adapt to, but most will be gradual rather than exceptional, cumulative rather than catastrophic, and many will be unlikely to be significantly out of the range we ordinarily experience through annual (or even day-on-day) fluctuations in weather. Even major flooding, as in 1953, has occurred in the recent past: it is only relative *risk* which will increase. Since individuals and social systems are particularly ill-suited to respond to gradual changes in their environments, it is not at all clear that short-term UK climate impacts, on their own, are likely to significantly shift public motivations, or to impact the lifestyle identities which currently drive unsustainable behaviours. And as discussed above.

The report ignores less directly observable, though probably equally important impacts such as species loss, changed land-use, or population migration.

The evidence is mixed on the impacts of extreme events such as flooding on climate beliefs. It remains to be seen, for example, whether the multiple UK and European flood events throughout the years 2000-2009 are now feeding through to more fundamental changes in attribution. The possibility for short-term change probably only exists where multiple visible events are also associated with development of consistent media and other discourses, which explicitly link these impacts to climate change and its anthropogenic causes - something which would also need to be expressed and communicated effectively in risk terms. Beyond 2050 (and under the 3-4 degree scenario) it becomes far more difficult to envisage a situation where climate change remains a hidden or secondary concern for most people. Equally, so many other factors may have changed by then so at to render predictions almost impossible.

Climate impacts occurring in more vulnerable locations across the globe over the next 40 years are, of course, likely to be of a different order to those occurring in the UK. But even these events might only serve to reinforce the existing belief that climate change is a matter for other countries and peoples, and for the UK just traditional disaster relief and emergency development aid. Major loss of life occurring in distant countries rarely generates the same level of emotional or cognitive engagement that individual, identifiable deaths do. Where external climate disasters do have a potential to shift beliefs, it is more likely to be with respect to adjusting society's long-term moral compass, towards incorporating a greater concern for fairness and equity for both other people and the environment. Expressing climate change to resonate with people's (moral) norms and values, even if this means transcending scientific uncertainty by couching it in terms of environmental, social and economic sustainability, is part of the challenge. In addition, technologies and approaches which genuinely embed within our everyday activities transparent linkages between behaviour and the environmental damage being caused will also have an important role to play. All of this suggests a clear need for effective communication allied to community and other forms of engagement around climate change objectives.

Some have argued that framing climate change in catastrophic, emotive terms may itself be counterproductive, by encouraging audiences to switch off or become habituated to the messages that they receive. However, while fear can lead to rejection and denial of messages, in other domains such as health protection fear communications are found to work in changing both attitudes and behaviour but only if they are accompanied by other factors such as perceived personal vulnerability, appropriate levels of social support, and above all practical actions that people can take to establish a level of control over the risk.

In summary, establishing climate salience is unlikely to come about in the near-term simply because of a recognition that physical climate impacts are directly occurring within the UK, even when some of these are severe. Achieving this would need a consistent framing, or narrative, linking climate risk, impacts and ongoing events. Such reframing is probably more likely to depend upon policymakers and stakeholders addressing factors other than the impact events themselves. Equally, other major global uncertainties which could impact before 2050 – conflict, continued economic recession, overpopulation and migration patterns, resource depletion and 'peak oil' – hold a far greater potential to influence future attitudes and behaviour towards the environment, climate and decarbonisation policies. This clearly cautions against basing forecasts about how society will evolve under the climate threat *solely* upon climate-oriented projections (e.g. the IPCC scenarios). For example, if depleting oil leads to sharply raised fossil fuel prices, pressures for radical energy conservation measures will also increase.

What major climate disasters are more likely to lead to is the opening up of temporary opportunities, or 'policy windows' as they have been termed by Kingdon, which would then allow legislators the license to take specific bold actions which they ordinarily believe would not otherwise be possible or politically acceptable. Past examples in the environmental and health arena include the introduction of the Clean Air Act after the London smog, and the development of the Thames Flood-Barrier after the 1953 East Anglia flooding. Of course disasters can also lead to ill-judged regulatory responses, but what Kingdon's model suggests is that there is merit in policy actors evolving *in advance* solutions to anticipated climate events (even though the political will to implement them might not currently be there). Risks, unfortunately, do bring policy opportunities, and the extent to which such windows for action might appear in the UK as a result of the occurrence of key climate-attributable events might form one line of socio-political analysis linked to the UK Climate Risk Assessment. In effect, envisaged solutions can become rapidly translated into practical options for action following a major disaster or near-miss. Radical change does depend upon the willingness of governments of the day (regional/local and national) to adapt policies in the light of the changed circumstances. Even here pressure for change may be limited to bolstering adaptation (e.g. better flood defences) rather than stronger mitigation to address underlying causes, given the close linkage between on-the-ground adaptation measures and demonstrable, local, human and environmental protection. What people are likely to care about most, in the aftermath of a climate-related disaster involving major loss of life or service disruption, would be any suggestion of government or corporate neglect in preparing for such events.

Perhaps better communication about climate change will impact beliefs and practices? An enduring assumption of much public policy in the environmental domain is that information provision is a sufficient condition for people to change their behaviours in an environmentally beneficial way. Risk and uncertainty framings are likely to be particularly important for this. While climate science and modelling has always acknowledged and incorporated complex uncertainties and gaps in its understanding, policy documents of the past have preferred to avoid risk. However, the 4th IPCC report in 2007 did adopt explicit (defined) likelihoods in describing scenarios and impacts, while the 2009 scenarios for the UK Climate Impacts Programme incorporate likelihood-based regional impact predictions to aid adaptation decision-making for the very first time. The UK Climate Risk assessment will, likewise, advance the policy and public debate about climate uncertainty around adaptation. Risk and uncertainty, then, are set to become part of the common currency underpinning debates about climate change decision-making. As a result, engaging the lay public about climate change will need to draw upon the very best guidance already developed within the field of communicating risk. These include such things as: avoid overly technical or patronizing language: choose risk terms with care, to contextualize numbers and risks in everyday terms, but without raising spurious comparisons; recognize diversity of audiences; avoid distrusted communication channels or parties; treat communication as dialogue (so as to learn as well as inform); combine information about harmful outcomes with actions which can help people to avoid the risk; and (always) evaluate communication impacts.

The reality of climate communication is complicated by cultural and political differences in beliefs - such that it is unwise to assume that there is any single public opinion on climate change just multiple 'publics'.

Communication design has to recognise this. There are also multiple barriers to change that mean that many people fail to act in a sustainable manner even when they accept a message. For example an individual may lack the funds to buy a more energy efficient product, may feel that they have other, more immediate priorities, or simply be reluctant to change their current habits and lifestyle. A useful distinction drawn within the health intervention literature is between upstream and downstream interventions. Downstream interventions refer to communications designed to change existing values and beliefs whilst upstream interventions refer to external structural changes including legal constraints and physical changes to the environment that themselves force, encourage, or more gently nudge people towards different practices and lifestyles. In reality a successful policy or campaign will often involve elements of both, and there is no reason to suspect climate change interventions will be any different in this regard.

The change in attitudes towards wearing seat belts, or against smoking in public places, for example, only fully came about after legislation first forced people to change their behaviour (coupled with long-term communication strategies to state why that change was necessary). What these and other examples suggest is: (a) that attitudes towards climate change, sustainable behaviour and individual responsibility for action on climate change more generally, can be influenced but only as part of well-designed structural interventions coupled with appropriately designed and evaluated (long-term) communications; and (b) public discord does not inevitably follow a change that legislators originally hesitate over, as beliefs and practices themselves often accommodate over time to externally imposed changes in behaviour. On this latter point we have also discussed evidence from different studies suggesting the conclusion that many people may currently be *ahead* of the political process in wanting bold, visible and demonstrably effective action on climate change from governments and other societal actors. Set against this, climate change and its causes are always likely to remain something of a hidden issue for most people in their day-to-day lives – and most of the time people will not particularly care about the precise details of climate policy as long as government actions are not seen as excessive or unfair. What may be more important in the future will be the engagement of the various engaged stakeholders - business, national and local authorities, NGOs and activist groups, media and scientists - in a way that effective climate policies can then be developed and fully implemented.

One exception to the above analysis may well be attitudes towards environmental taxation, where there tends to be less support for measures which impact people personally, than for other - e.g. technological – options, particularly amongst those on the political right. The UK transport fuel protests in 2000 illustrated how disruptive action can be rapidly mobilized around opposition to environmental taxation. In that example, however, it was clear that successive governments, Conservative and Labour, had failed to properly explain to people why the fuel escalator was in place. Indeed, much of the current discourse surrounding climate change denial is bound up with wider beliefs about opposition to state intervention and personal taxation. The extent to which measures are framed in this way or in alternative terms (e.g. as contributing towards dedicated decarbonisation policies, or as part of a climate compensation fund), and more critically the extent to which they are perceived to be fair, are all likely to impact upon the ways in which they are received. It is also not at all clear from the literature – in part because this issue is not well studied – whether members of the public grasp the full scale of changes necessary if the UK is to meet its tough emissions targets by 2050. When asked what they are doing about climate change the answer is often 'recycling', and recycling rates have indeed increased dramatically across the UK in the past 10 years. While worthwhile activity in and of itself, recycling is not necessarily the best way of effecting major reductions in our personal carbon emissions. It may well be that a radical reframing of public opinion would come about if the true nature and scale of the problem were made clearer to people.

By contrast to climate impacts and direct communications, powerful contextual factors shaping our behaviour, practices and attitudes towards the environment and climate change are far more likely to arise from future societal responses to the climate threat. These will include system-wide changes to major energy infrastructure and technology, availability of more sustainable and energy efficient goods and services, changes to transport infrastructure, rising energy prices, and various adaptation measures. All of these hold the potential to change not only people's behaviour, practices and lifestyles (some quite quickly) but with this also beliefs about climate change. Indeed, to work as intended many of the planned new technologies and low carbon interventions will *depend* upon people changing their behaviour and beliefs, whether this is through uptake of home efficiency programmes, use of electric vehicles and 'eco-driving', the acceptability of new nuclear power or onshore wind, or the acceptance of new fiscal regimes.

One area of energy and climate policy where we do have considerable empirical experience concerns public attitudes to energy production. Energy system change bears upon multiple long-term national policy goals, including the transition to a low carbon economy, energy security and affordability, and mitigating wider environmental impacts. Significant interrelated transformations in the way the UK supplies, manages and consumes its energy will be essential if these aims are to be attained. Public values, attitudes and acceptability will be of critical importance in these processes of whole energy systems transformation, with the potential to present both opportunities and challenges for the delivery of energy policy and change across multiple areas. Demand side shifts will involve behaviour change to decrease energy consumption as well as the uptake of new energy management technologies such as interactive smart metering. On the supply side considerations relate to major infrastructure siting (e.g. acceptability, equity) and the lifestyle changes implied in new supply configurations (e.g. an all electric future). We have considerable understanding of public (un)acceptability of single technologies such as wind or nuclear power, and insights into individual behaviour-change options.

However, there is scant evidence regarding potential public responses to the sorts of whole energy-system transformations (e.g. to all electric homes or transportation) which are now being discussed as serious options for 2050 and beyond. Public attitudes towards innovative technologies, as well as towards whole energy-system change, are therefore important areas for future research. In a similar fashion, we can only speculate at this stage how some of the more radical technological proposals for addressing climate change, such as climate geoengineering, will interact with public views if they are ultimately developed and deployed. Will such technologies prove a wake-up call for people, leading to recognition of the scale and seriousness of the problem, or will they lead to a situation insurers describe as 'moral hazard', where people believe that the pressure to mitigate carbon emissions is reduced because of the promise of an easy technological solution? Again there is a need for more detailed empirical work on this issue.

In their thoughtful analysis of why most countries have not moved further down the route of deep emission cuts, despite the overwhelming scientific evidence of climate change, Compston and Bailey argue that the failure is essentially a political one. They cite six factors underlying this: the perception that individual countries can do little on their own; the influence of climate sceptics; a shortage of viable technical and economic solutions; the problem of economic competitiveness between nations; fear of electoral retribution; and structural obstacles within government. Based upon their analysis it would appear that the greatest challenge for many countries around the world in the face of the growing threat of climate change will be one of governance, rather than merely of individual attitudes and behaviour. The UK has begun to address some of these political barriers through moves to reach cross-party consensus on climate targets in Parliament, the Climate Change Act and creation of the Climate Change Committee, and by placing energy and climate policy under the same ministerial portfolio. However, more does need to be done, and substantial political and economic barriers in the UK remain (including, for example, the separation of adaptation and mitigation policy), to which could be added a failure of the theory of ecological modernisation – the somewhat questionable assumption that growth in the global economic system would automatically be accompanied by better levels of environmental protection. The economist Tim Jackson has argued that the financial and ecological crises share the same underlying systemic origins, and that in order to face up to our major environmental and sustainability challenges we need a new economic paradigm stressing 'prosperity without growth'. As a final remark, one could add that we also need to devise a democracy which votes for uncertain, but possibly vital future pathways that will maintain a sustainable planet for all. Achieving such a democracy will have to involve a genuine dialogue with electors and civil society more widely, on terms that neither patronises them nor minimises the considerable stakes involved.

References

Giddens, A. (2009) *The Politics of Climate Change*. Polity Press

Hulme, M. (2009) *Why We Disagree About Climate Change*. Cambridge: Cambridge University Press.

Kempton, W. (1991) Public understanding of global warming. *Society and Natural Resources*, 4(4), 331-345.

Leiserowitz, A. (2007a) *Public Perceptions, Opinion and Understanding of Climate Change – Current Paterns, Trends and Limitations.* United Nations Development Program, Human Development Report Office, Occasional Paper (no 2007/31).

Lorenzoni, I. and Pidgeon, N.F (2006) Public views on climate change: European and USA perspectives. Climatic Change, 77, 73-95.

5

Vulnerability of People and the Environment

There are strong causal relationships among the state of the environment, human well-being and vulnerability. Understanding how environmental and non-environmental changes affect human wellbeing and vulnerability is the critical foundation for addressing challenges to and the opportunities for improving human well-being while also protecting the environment. Vulnerabilities are often driven by actions taken at a great distance, highlighting worldwide interdependencies. This analysis also evaluates whether environmental governance adequately links with other relevant policy domains, such as poverty alleviation, health, science, and technology and trade. It underlines the need for mainstreaming environment into these domains to reduce vulnerability. This provides strategic directions for policy making to reduce vulnerability and enhance human well-being.

As the World Commission on Environment and Development (Brundtland Commission) stated in *Our Common Future*, "A more careful and sensitive consideration of their (vulnerable groups) interests is a touchstone of sustainable development policy". The vulnerability approach applied here shows the potential for strong negative consequences for well-being of, for example, reduced access to resources, such as food and drinking water, and the existence of thresholds beyond which health and survival are severely threatened. Patterns of vulnerability to environmental and socio-economic changes, here referred to as "archetypes," describe the impacts of these changes on human well-being.

Global Context of Vulnerability

A number of factors shape the vulnerability of people and the environment, including poverty, health, globalization, trade and aid, conflict, changing levels of governance, and science and technology.

Poverty

Poverty reduces the ability of individuals to respond and adapt to environmental change. Although the multidimensional nature of poverty is widely recognized, income and consumption remain the most common measures. Most regions have made progress in meeting the first Millennium Development Goal (MDG 1) on reducing extreme poverty and hunger, although many will not achieve the 2015 targets. Actual numbers decreased from 1.2 billion to just over 1 billion in 2002. The percentage of people in the world with insufficient food to meet their daily needs has declined, but actual numbers increased between 1995 and 2003, when about 824 million people suffered chronic hunger. Sustained growth in China and India has contributed to sharp decreases in extreme poverty in Asia. Where inequity is high, including in some of the transition countries of Europe and Central Asia, economic growth does not necessarily translate into less poverty. In many countries, relative poverty is increasing despite general affluence. In the United States, for example, the number of people living below the national poverty line has risen since 2000, reaching almost 36 million in 2003. Structural economic adjustment, ill health, and poor governance affected progress in some regions, including sub-Saharan Africa.

Health

Health is central to the achievement of the MDGs because it is the basis for job productivity, the capacity to learn, and the capability to grow intellectually, physically and emotionally. Health and education are the cornerstones of human capital. Ill health reduces the capacity to adapt to environmental and other changes. Under-five mortality rates have improved considerably, though there are still large regional differences, and more than 10 million children under five still die every year – 98 per cent of them in developing countries. Some 3 million die due to unhealthy environments.

They include traditional risks associated with underdevelopment (such as underweight, unsafe water and lack of sanitation), and those associated with consumptive lifestyles (such as obesity and physical inactivity).

Health gains are unequal across regions and within population groups. In the least favourable health situations, people suffer persistent communicable diseases associated with deficient living conditions, including poverty and progressive environmental degradation. AIDS has become a leading cause of premature deaths in sub-Saharan Africa, and the fourth largest killer worldwide. By the end of 2004, an estimated 39 million people were living with HIV/AIDS. The epidemic has reversed decades of development progress in the worst-affected countries, contributing to strong increases in vulnerability.

Globalization, trade and aid The rapid growth of trade and financial flows is creating more global interdependence. The trade and development agendas have so far not been reconciled, and the gulf between the rich and the poor is still growing. Poor countries are moving to market solutions and pragmatic arrangements for increasing trade and foreign direct investment (FDI) to create more jobs and alleviate poverty. The outcomes are highly uneven. The failure of the Doha round of the WTO talks continues to hurt the poorest of the poor who often depend on agricultural markets.

Table 1. Estimated attributable and avoidable burdens of 10 leading selected risk factors

Developing countries high mortality (per cent)		*Developing countries low mortality (per cent)*		*Developed countries (per cent)*	
Underweight	14.9	Alcohol	6.2	Tobacco	12.2
Unsafe sex	10.2	Blood pressure	5.0	Blood pressure	10.9
Unsafe water, sanitation and hygiene	5.5	Tobacco	4.0	Alcohol	9.2
Indoor smoke from so id fuel	3.6	Underweight	3.1	Cholesterol	7.6
Zinc deficiency	3.2	Overweight	2.4	Overweight	7.4
ron deficiency	3.1	Cholesterol	2.1	Low fruit and vegetable intake	3.9
Vitamin A deficiency	3.0	Low fruit and vegetable intake	1.9	Physica inactivity	3.3
Blood pressure	2.5	Indoor smoke from so id fuel	1.9	Illicit drugs	1.8
Tobacco	2.0	ron deficiency	1.8	Unsafe sex	0.8
Cholesterol	1.9	Unsafe water, sanitation and hygiene	1.8	ron deficiency	0.7

With the growing interest in markets, the aid agenda has also changed. Most of the recent increases in aid have been used to cancel debt, and meet humanitarian and reconstruction needs following emergencies. The share of total official development assistance (ODA) going to basic human needs has doubled since the mid-1990s, but the share going to agriculture and physical infrastructure has diminished. These two sectors need support if countries are to feed their own people, build their economies, and increase their adaptive capacity. Africa remains the most aid dependent region by far, while West Asia's dependence on aid has varied considerably over the past 20 years. Together the figures suggest a bleak reality. FDI, which is productive capital, is a great deal lower than aid in many regions. In 2005, the 191 million migrants worldwide contributed in excess of US$233 billion to productive capital of which US$167 billion went to developing countries.

Conflict

The end of the Cold War in the late 1980s has reduced the threat of nuclear war from great power rivalry, although fears of continued nuclear proliferation among states and non-state actors remain. Civil conflicts continue to be the biggest threat, although incidences have decreased dramatically in recent years. International involvement in civil wars, primarily in peacemaking and peacekeeping capacities, is at an all-time high due to humanitarian pressures. The increase in the number of formal democracies is unprecedented; this trend may contribute to the decreasing incidence of civil wars, although the transition to democracy is often a highly unstable period. All regions of the world have seen a decrease in armed violence except for sub-Saharan Africa and West Asia.

Despite the positive global trends in armed violence, persistent conflicts have very negative impacts on well-being. More than 8 million people have died directly or indirectly as a result of war in Africa since 1960. Conflicts, violence and fear of persecution regularly displace large civilian populations, forcing millions of people into marginal ecological and economic areas within countries and across international boundaries. The UNHCR estimates that there were 11.5 million refugees, asylum seekers and stateless persons, and 6.6 million internally displaced persons globally in 2005. The forced movement of people into marginal areas undercuts, sometimes for decades, sustainable livelihoods, economic development, and societal and state capacities. The resulting poverty, often tied to shortages or degradation of natural resources, contributes directly to lower levels of well-being and higher levels of vulnerability.

Changing levels of governance Over the past 20 years, governance has become increasingly multi-level, with more interaction and interdependence between different levels. The effectiveness of national policies remains mixed, but the capacity and political will of governments to take action has increased. In combination, these trends increase opportunities to reduce vulnerability. The early years after the end of the Cold War witnessed a renewed optimism in multilateralism and global governance. In parallel, regional cooperation made significant progress around the world, even if its forms and intensity differ.

There has also been a trend towards political and fiscal decentralization from national to sub-national levels, including in countries of the Organisation for Economic Cooperation and Development (OECD) and in Africa and Latin America. This may not necessarily mean that local authorities have been empowered, as decentralization without devolution of power can be a way to strengthen the presence of the central authority. Local governments, community-based groups and other non-governmental actors now engage more widely in international cooperation, contributing to a better grounding of global policy in experiences of local vulnerability. Global corporations' influence has extended beyond the economic arena (De Grauwe and Camerman 2003, Graham 2000, Wolf 2004), and many choose to develop voluntary environmental codes, and to increase self regulation.

Science and Technology

Developments in science and technology have helped reduce human vulnerability to environmental and non-environmental change, although the pace and levels at which different regions achieve progress vary widely. Expenditures on research and development in OECD countries between 1997 and 2002 were 2.5 per cent of GDP compared to 0.9 per cent of GDP in developing countries. While the number of researchers was 3 046 per million people in OECD countries between 1990 and 2003, it was 400 per million in developing countries. The potential for science and technology to reduce vulnerability remains very unevenly distributed worldwide. This illustrates the need to improve technology transfers between regions.

For example, new farming technologies and practices since 1960 increased food production, and decreased food prices, addressing undernutrition and chronic famine in many regions, but access to these technologies remains unevenly distributed. In the 1980s, oral rehydration therapies and vaccines suitable for use in developing countries were critical in reducing under-five mortality. New information and communication technologies give unprecedented opportunities for early warning systems, and for generating local entrepreneurship. However, science and technology have undoubtedly also added to the risks faced by people and the environment, particularly by driving environmental change.

Human Well-being, Environment and Vulnerability

Development Challenges

Improving human well-being – the extent to which individuals have the ability to live the kind of lives they value, and the opportunities they have to achieve their potential – is at the heart of development. This is not just a moral imperative, but also a critical aspect of fundamental human rights, and is essential for reducing vulnerability and achieving sustainable use of the environment.

Since the 1987 Brundtland Commission report emphasized the environment-development link, different policy statements and multilateral environmental agreements, including the 1992 Rio Declaration (Principle 1) and the conventions on biological diversity and climate change, have highlighted the opportunities the environment holds for development. Increased convergence between these international approaches and those at national level is evident from the highest-level recognition of environmental rights as human rights (Ncube and others 1996, Mollo and others 2005). Importantly, environmental rights approaches have moved from a focus on environmental quality to incorporating basic needs, development, and intergenerational and governance concerns (UN 2003, Gleick 1999, Mollo and others 2005). However, progress in meeting development objectives has been uneven.

Improvements in well-being – for some Despite significant improvements in well-being over the last 20 years, with gains in income, nutrition, health, governance and peace, there are many ongoing challenges. Millions of people across all regions are poor, and lacking essential services that are now common among the wealthy. Many countries will not meet the MDGs' 2015 targets. But the environment provides opportunities to meet these goals, and to enhance well-being through the various goods-and-services it provides.

The link between environment and well-being is complex, non-linear and influenced by multiple factors, including poverty, trade, technology, gender and other social relations, governance, and the different aspects of vulnerability. Global interconnectedness – through a shared natural environment and globalization – means that achieving human well-being in one place may be affected by practices elsewhere.How people actually live and the opportunities they have are closely connected to the environment. As the Brundtland Commission warned, environmental degradation contributes to "the downward spiral of poverty" and amounts to "a waste of opportunities and of resources". Good health, for example, is directly dependent on good environmental quality. Many national constitutions now recognize a healthy environment as a fundamental human right. Despite some improvements, pollution continues to be a problem, sometimes spurred on by factors outside the control of its victims. Associated risks and costs are unevenly distributed across society. Although the incidence of ill health has been reduced globally, the costs remain monumental.

Notwithstanding improvements in access to water and sanitation, the poorest people suffer the greatest water deficit as a result of location, poor infrastructure and lack of financial resources. Consequently, they experience ill health and indignity. In many developing countries, poor people in cities pay more for water than wealthier inhabitants. Poor access to material assets at the household level (income, food, water, shelter, clothing, energy, natural and financial resources) and at the societal level (physical and service infrastructure) is part of a cycle of impoverishment, vulnerability and environmental change. It is part of a sequence of becoming poor and staying poor. In developed countries too, relative poverty, age and gender are critical factors in the distribution of benefits. The energy archetype illustrates the vulnerabilities that arise through lack of access to energy, as well as those related to dependency on energy imports. Investing in physical and service infrastructural development can improve well-being by increasing marketing opportunities, security, and access to energy, clean water and technologies for efficient and sustainable natural resources use.

In countries with a low human development index, people also live shorter lives, because they have reduced health, due to hunger, unsafe water, sanitation and hygiene, lead exposure, and climate change. Gains in life expectancy, child mortality and per capita health expenditures have been systematically greater in those countries with more equitable income distribution and access to medical treatment. Costa Rica, for example, has a higher average life expectancy than the United States. In many wealthier societies, opulence and consumerism, as well as relative poverty, contribute to ill health.

Investing in Human and Social Capital Reduces Vulnerability

Environmental assets can provide important opportunities for improving well-being but, as shown in the archetypes, too often the benefits from these resources do not reach the most vulnerable. The distribution of environmental benefits is affected by access to networks (for example NGOs, governments and the private sector) and relations of trust, reciprocity and exchange. Development processes that arbitrarily extinguish local rights and degrade the environment, as well as global trade regimes are also important factors influencing distribution.

Several policy interventions respond to these challenges, but slow progress in achieving the MDGs in many countries suggests that not enough has been done. The Convention on Biological Diversity (CBD), for example, emphasizes the importance of more equitable sharing of conservation benefits. Agenda 21, the Rio Declaration, and the CBD all prioritize public participation as essential for sustainable development. Increasing income from benefit sharing may strengthen efforts to meet MDG 1, and as household resources increase, the education and health-related MDGs may be more achievable. Countries with low access to improved drinking water have lower equity in access to education. Worldwide, girls and women spend about 40 billion hours collecting water – equivalent to a year's labour for the entire workforce in France. In many developing countries, women and girls spend more than 2 hours a day collecting water. There are strong positive linkages between progress on the different MDGs, with, for example, improved access to water (MDG 7), resulting in girls spending less time collecting water, and increasing their opportunities to attend school. For many countries, effectively implementing an interlinkages approach is challenging.

Meeting basic needs, such as education and health, provides the basis for valued choices, and enhances the day-to-day capacity of individuals, including that for environmental management. Education and access to technology are particularly important in poor communities, where they provide a potential route to a better situation and reduced vulnerability.

Basic capabilities and rights to be treated with dignity, to have access to information, to be consulted and to be able to give prior informed consent where one's livelihood or assets are affected, are increasingly recognized as social and economic rights. The 1986 UN Declaration on the Right to Development represents a global consensus, but for many, these rights are inaccessible as a result of weak national and regional governance systems, undercutting capacity and opportunities. Women remain particularly disadvantaged. Notwithstanding improvements in maternal health (MDG 5) resulting, for example, from improved access to technologies and energy in rural hospitals, and access to education (MDG 3) in all regions since 1990, women continue to be among the most disadvantaged. They are under-represented in the economy and decision making.

Women are under-represented in important parts of society, due to a combination of factors. Socio-cultural attitudes, education, employment policies, and a lack of options for balancing work and family responsibilities and for family planning affect opportunities for employment and participation in community affairs.

Personal security – being protected from or not exposed to danger, and the ability to live a life one values – may be threatened by declining social cohesion, poor living standards, inequity, unfair distribution of benefits and environmental change. In some circumstances, environmental change creates a security challenge for entire cultures, communities, countries or regions. Where (cultural) identities are closely associated with natural resources, as in the Arctic and many Small Island Developing States (SIDS), social conflict and breakdown may be directly linked to habitat destruction or decreasing availability of environmental services. Other contributing factors include low levels of rural growth, high income inequity, ill health (especially HIV prevalence), climatic factors, such as drought, and environmental degradation.

Conflict also affects food security because of its long-lasting disruption of the productive base, and its impact on overall human well-being. In many cases, countries involved in conflict, and those with high levels of inequity, experience higher than expected levels of food emergencies.

Investing in good social relations, building social capital through better governance, improving cooperation, and empowering women not only supports conservation efforts, but builds opportunities for peace, development and improving well-being. Developed countries' experiences suggest a number of factors that help hedge the impact of disasters: a well-financed government, an insurance industry, transport and communication infrastructure, democratic participation and personal affluence. Improving capacity and access to technology, as envisaged under the Johannesburg Plan of Implementation (JPOI) and the Bali Strategic Plan for Technology Support and Capacity Building (BSP), can improve coping capacity. However, progress towards developing the global partnership to support this access remains slow. More far-sighted and equitable approaches to the movement of resources, goods and people are critical to address the new levels of stress the most vulnerable communities will face as a result of environmental change.

Aspects of Vulnerability

Although vulnerability is context and site specific, certain common elements can be observed across various regions, scales and contexts. Overarching vulnerability issues, such as equity, the export and import of vulnerability from one place or generation to another, and the causal relationships with conflict, hazards and the environment, deserve special attention, since they represent strategic entry points for effective vulnerability reduction and policy making.

Inequalities, equity and vulnerable groups. Vulnerability varies across categories, including among men and women, poor and rich, and rural and urban, as can be observed in all archetypes. Refugees, migrants, displaced groups, the poor, the very young and old, women and children are often the groups most vulnerable to multiple stresses. Factors such as ethnicity, caste, gender, financial status or geographical location underlie processes of marginalization and disempowerment, which all lower the capacity to respond to changes. For example, the access of women and children to health care is often inequitably distributed, resulting in unfair and unjust outcomes that entrench disadvantage. Gender inequalities, reflected, for example, in male and female differences in wages, nutrition and participation in social choice, are illustrated in the contaminated sites archetype. Addressing MDG 3, to promote gender equality, empower women and eliminate gender disparity in primary and secondary education, is essential for increasing women's opportunities, reducing their vulnerability, and improving their ability to create sustainable and sufficient livelihoods.

One response by communities and governments to the unequal distribution of vulnerability and the impacts of multiple stressors on human well-being has been to focus on issues of environmental justice.

Export and import of vulnerability. Vulnerability is created or increased remotely, in many cases, through cause-and-effect relationships that persist over long distances in space or time. Many vulnerability archetypes demonstrate the phenomenon of "vulnerability export." Decreasing the vulnerability of some, for example through provision of shelter, increases the vulnerability of others far away, for example through land degradation and contamination around mining areas for building materials. At the same time, many people in industrialized nations, and the new consumers in the developing countries do not feel most of the impacts on the environment that result from their behaviour. These negative effects on the environment and well-being (especially health, security and material assets) are felt most strongly by those, especially the poor, living where the resources are extracted or the waste is dumped.

The emissions and land degradation associated with extraction and processing of the materials are increasing in developing countries, while the high-value end products are consumed in industrialized countries. Similarly, food imports often mean that environmental degradation and social impacts occur in the producing land, rather than where the goods are consumed. Vulnerability is imported where, for example, there is agreement to import waste and hazardous materials to locations where it cannot be safely disposed of or managed. The vulnerability of local populations is created or reinforced by poor governance and a lack of capacity to deal with the hazardous materials. Inadequate storage and poor stock management often result from insufficient storage capacity for pesticides, inappropriate storage conditions, insufficient training of responsible staff in stock management, poor distribution systems, inappropriate handling during transport, and unavailability of analytical facilities.

While international trade can lead to increases in income, and has helped millions of people out of poverty, it is also sustaining unequal patterns of consumption, and in outsourcing the extraction of natural resources, much of the production and manufacturing, and also the generation and disposal of their hazardous wastes.Recently, however, there have been some attempts to include the external impacts of trade policies into decision making processes, for example, through sustainability impact assessments in the European Union.

Vulnerability, environment and conflict. Many of the patterns of vulnerability represent a potential for or have already led to conflict. The relationship between environmental problems and international and civil conflict has been the subject of a great deal of academic research in the post-Cold War period. Both scarcity and abundance of environmental resources can exacerbate existing tensions, and contribute to conflict between groups, especially in societies that lack the capacity to effectively and equitably manage competition for control over resources. These dynamics tend to be most common in the developing world. However, the export of vulnerability from developed to developing countries, can mean that even conflicts that appear localized have critical external connections.

A combination of environmental change, resource capture and population growth decreases the per capita availability of natural resources, and can threaten well-being for large segments of societies, particularly the poorest who depend on these natural resources for survival. The resulting social effects – migration, intensified unsustainable behaviour and social sub-grouping – strain the state's ability to meet its citizens' demands, and can contribute to violent outcomes. In the dryland archetype, conflict potential is related to unequal access to scarce water, forest and land resources, exacerbated by desertification and climate variability. Migration, a traditional coping strategy, sometimes heightens conflict when migrants create new competition for resources, or upset tenuous cultural, economic or political balances in the receiving area. In other cases, the scarcities heighten tensions between nomadic and pastoralist communities. Where this migration occurs across international boundaries, it can contribute to interstate tension and new civil strife. Even when a state's natural resource base is high, conflict can erupt over control of these valuable resources, if the potential cost of waging war is lower than the potential gains associated with securing access to the resources for export.

In the archetype on technological approaches to water problems, conflicts and tensions surrounding the distribution, access and quality of water resources arise. Megaprojects, such as dam construction, often carry considerable costs, including forced displacement for riparian dwellers, who may receive few of the resulting benefits. These costs may include tensions between the state and riparian users, as well as between upstream and downstream riparian groups. The overexploitation of global commons, such as fisheries, the focus of another archetype, brings smaller-scale fisher groups and their governments into conflict with transnational or foreign-flagged ships that venture into exclusive economic zones from the depleted commons. Future energy generation and climate change directly link to security concerns for both oil-importing and oil-exporting countries. In rapidly urbanizing coastal zones and SIDS, conflicts emerge over competition for the environment for tourism-related activities, or for its environmental services associated with marine ecosystems and local livelihoods. Greater attention to proper management of ecosystems and valuable resources promises lower vulnerability to violence and greater overall well-being.

Vulnerability, well-being and natural hazards. Over the past 20 years, natural disasters have claimed more than 1.5 million lives, and affected more than 200 million people annually. One of the main drivers of increased vulnerability to hazards is global environmental change. Natural hazards, such as earthquakes, floods, droughts, storms, tropical cyclones and hurricanes, wildfires, tsunamis, volcanic eruptions and landslides threaten everyone. Proportionally, however, they hurt the poor most of all. Global datasets on extreme events indicate that the number of natural hazards is increasing. Two-thirds of all disasters are hydrometeorological events, such as floods, windstorms and extreme temperatures.

The consequences of disasters can have a lasting impact, threaten achievements in development and undermine resilience. Natural hazards affect food security, water supply, health, income and shelter. These impacts are illustrated in several of the archetypes. Insecurity is driven by a multiplicity of environmental, political, social and economic factors, and is also closely related to issues of material access and social relations. Inefficient and poor governance, as well as inadequate or inefficient early warning and response systems, exacerbate vulnerability and the risks associated with environmental change and natural disasters. In some cases, short-term disaster relief even contributes to increasing long-term vulnerability.

Exposure to hazards has increased as a result of climate change and, for example, destruction of mangroves that protect coasts from tidal surges, but also through the continuing concentration of population in highly exposed areas. Adaptive capacity is also being eroded through, for example, reduced state social protection schemes, undermining of informal safety nets, poorly built or maintained infrastructure, chronic illness and conflict.

Patterns of Vulnerability

Recurring patterns of vulnerability can be found in numerous places around the world, including industrialized and developing regions, and urban and rural areas. With the recognition of the relevance of multiple pressures, and the close interlinkages among local, regional and global scales, vulnerability analyses become increasingly complex. For detailed local vulnerability case studies, there is the question of their relevance for other parts of the world, but it is possible to recognize some similarities between cases and to draw policy-relevant lessons from them.

An archetype of vulnerability is defined as a specific, representative pattern of the interactions between environmental change and human well-being. They do not describe one specific situation, but rather focus on the most important common properties of a multitude of cases that are "archetypical." The approach is inspired by the syndrome approach, which looks at non-sustainable patterns of interaction between people and the environment, and unveils the dynamics behind them. The archetype approach is broader, as it includes opportunities offered by the environment to reduce vulnerability and improve human well-being.

The archetypes presented here are simplifications of real cases, to show the basic processes whereby vulnerability is produced within a context of multiple pressures. This may allow policy-makers to recognize their particular situations in a broader context, providing regional perspectives and important connections between regions and the global context, and insights into possible solutions. The patterns of vulnerability are not mutually exclusive. In some ecosystems, countries, sub-regions, regions and globally, a mosaic of these and other patterns of vulnerability may exist. This makes policy response a complex challenge.

The archetypes of vulnerability have been identified through the GEO-4 assessment, ensuring regional relevance and balance. The seven archetypes presented here are not meant to provide an exhaustive overview of all possible patterns of vulnerability. However, they provide a good basis for identifying challenges and exploring opportunities for reducing vulnerability while protecting the environment.

Exposing People and the Environment to Contaminants

The archetype concerns sites at which harmful and toxic substances occur at concentrations:

- above background levels and pose or are likely to pose an immediate or long-term hazard to human health or the environment; or
- exceed levels specified in policies and/or regulations.

People and ecosystems are exposed to widespread contamination due to persistent organic pollutants and heavy metals, urban and industrial sites, military activity, agro-chemical stockpiles, leaking oil pipelines and waste dumps.

Much work is still needed to quantify the extent of contamination due to toxic and hazardous substances, and to make governments and civil society aware of the problems. However, a considerable amount of contamination has been documented.

In addition to contamination generated in particular locations, transport and deposition of waste is a major threat. More than 300 million tonnes of waste, including hazardous and other wastes, were generated worldwide in 2000, of which less than 2 per cent was exported. About 90 per cent of the exported waste was classified as hazardous, with about 30 per cent believed to be persistent organic pollutants (POPs). Contaminated sites are also legacies of past industrial and economic development, and a heritage of present production and consumption patterns that affect both current and future generations. Abandoned industrial sites can present a serious risk to people and the environment. Governments face problems of holding polluters accountable for site clean-ups. Therefore, clean-up costs are imposed on state budgets, or on people from surrounding areas exposed to health risks and environmental deterioration.

Sometimes, abandoned industrial sites are in relatively isolated areas around former factories or mines, and, sometimes, whole regions are affected by the problem. Short-term profit interests, lack of regulations or corruption, and weak law enforcement are among the factors that have led and may still lead to the creation of present and future environmental hazards from contaminated sites.

Vulnerability and human well-being. In developing countries, chemical mixtures in the vicinity of small-scale enterprises, such as smelters, mines, agricultural areas and toxic waste disposal sites, are often a human health hazard. For example, about 60 per cent of the smelters of the world are located in developing countries, while developed countries import the metals. Health effects, such as cancer and neuropsychological disorders, have also been reported around smelters. For example, in Torreon, Mexico, 77 per cent of the children living closest to a lead smelter had lead levels twice as high as the reference level.

Mercury contamination associated with small-scale gold mining and processing presents a major hazard for environment and human health in at least 25 countries in Africa, Asia and the Pacific, and Latin America and the Caribbean, Harmful health effects have been reported for individuals exposed to mercury in gold mining areas.

Pesticides can contribute to water pollution, and seriously threaten the health of both rural and urban residents, especially the poorest people. Organochlorine compounds, such as DDT, dieldrin and HCH, which have been withdrawn or banned for human health and/or environmental reasons, are still found in dumps, particularly in developing countries. Long-term exposure to pesticides can increase the risk of developmental and reproductive disorders, disruption of the immune and endocrine systems, and can impair the function of the nervous system, and is associated with the development of certain cancers. Children are at higher risk from exposure than are adults. The international traffic in hazardous wastes exposes local populations to health risks. For example, in 1998, about 2 700 tonnes of industrial waste, containing high levels of toxic compounds such as mercury and other heavy metals, were shipped illegally to Sihanoukville, Cambodia. An estimated 2 000 residents were exposed to the waste and at least six deaths and hundreds of injuries were associated with the incident.

An emerging issue is the great volume of electronic waste exported to developing countries, where it is recycled by workers who often lack protection. They are exposed to mercury, lead, cadmium and other toxic chemicals. In one Chinese city where electronic waste is recycled, sediment samples had heavy metal concentrations far above the guidelines of the US Environmental Protection Agency. Similarly, workers are exposed to contaminants that pose serious risks to their health in locations where ships are broken up for recycling.

Abandoned factories and industrial sites are most likely to be found in poor communities, which can be home to marginalized newcomers. Contamination of air, water and land decreases land productivity, making agricultural products unsuitable for markets. Children are particularly at risk from contaminated sites (as places of play and work), and women are especially at risk for physiological reasons. A survey conducted in the United Kingdom about the social status of people living close to integrated pollution control sites (IPC), confirmed that in England there is strong evidence of a socially unequal distribution of IPC sites and their associated potential impacts. Out of about 3.6 million people living in one-kilometre radius of an IPC site, there were six times more people from the most deprived groups than from the least deprived groups.

Responses

Over the years, a series of measures have been adopted to deal with the risks that hazardous materials and chemicals pose for both people and the environment. Principle 14 of the Rio Declaration, calls on countries to "effectively cooperate to discourage or prevent the relocation and transfer to other States of any activities and substances that cause severe environmental degradation or are found to be harmful to human health." The UN Commission on Human Rights has appointed a special rapporteur on adverse effects of the illicit movement and dumping of toxic and dangerous products and wastes on the enjoyment of human rights (UN).

Responses to the problem of contaminants now include 17 multilateral agreements, together with numerous intergovernmental organizations and coordination mechanisms. They include the 1989 Basel Convention on the Control of Transboundary Movements of Hazardous Waste and Their Disposal, the 1998 Rotterdam Convention on Prior Informed Consent Procedure for Certain Hazardous Chemicals, the 2001 Stockholm Convention on Persistent Organic Pollutants, as well as the 2006 Strategic Approach to International Chemicals Management.

Other responses to contamination have created opportunities for building trust in post-conflict societies. For example, joint scientific assessment of threats from radioactive contamination in the Russian northwest provided an opportunity for Russian, Norwegian, and American exchange as the Cold War ended and the superpowers began to develop inks for confidence building among scientists and military personnel. The low politicization of environmental issues actually facilitated face-to-face dialogue among military foes in a highly militarized and sensitive region.

The success of the existing instruments for dealing with contamination depends strongly on institutiona capacity and political will. Important areas for future action include:

- strengthening the ability of internationa organizations to monitor and enforce multilatera agreements, such as the Basel and Rotterdam conventions;
- promoting global environmental and socia standards to avoid dumping;
- investing in technology and technology transfer for improved risk assessment, monitoring, information and communication, and clean-up;
- increasing corporate social and environmenta responsibility;
- investing in assets, especially skills and knowledge, to avoid exposure or to mitigate health effects from exposure to hazardous material;
- improving state capacity to monitor and enforce laws, as this may reduce risk, and improve local coping capacity;
- providing opportunities for participation, and addressing the social situation of people affected by contaminated sites;
- better incorporation of established internationa legal principles - including the precautionary approach, producer liability, polluter pays, prior informed consent and right to know - into national, regional and global frameworks;
- increasing support for research on causes and effects (especially cumulative effects) of industrial production and chemicals; and
- increasing support for life cycle analyses and environmental impact assessments.

In situations of contaminated sites, formal institutions, better laws at national and international levels, and better enforcement of existing laws are crucial for reducing vulnerability. This requires strong and well-functioning states, with law-making, implementation and enforcement branches working towards the same goals. Measures that strengthen the capacity of states can also help strengthen coping capacity at local levels, if this is supported by higher levels of governance. Increasing the participation of the most vulnerable groups in planning and governance, and giving both local and higher levels of governance opportunities to articulate their challenges is a major factor in strengthening their coping capacity. Giving the vulnerable a voice requires that they be actively empowered to raise their voices, for example by having access to relevant environmental information – as enshrined in Principle 10 of the Rio Declaration – and capacity building for taking part in the governance process. The 1992 UN Conference on Environment and Development (UNCED) provided the basic institutional change for increasing participation in environment-related decision making. This has been reinforced, for example, in the Aarhus Convention. The Basel and Rotterdam conventions are important for giving countries a voice in the context of vulnerability to contamination.

Disturbing the fragile equilibrium in drylands

In this archetype, current production and consumption patterns (from global to local levels) disturb the fragile equilibrium of human-environment interactions that have developed in drylands, involving sensitivity to variable water supplies and resilience to aridity. The result is new levels of vulnerability. For thousands of years, drylands populations have been dependent on the proper functioning of these ecosystems for their livelihoods. These resilient ecosystems have considerable productive potential – supporting, for example, 50 per cent of the world's livestock – but are increasingly at risk. Moreover, governance and trade patterns mean that much dryland wealth remains hidden or poorly used, constituting missed opportunities for improving well-being.

Drylands are widespread, occur in developed and developing countries, and support significant populations. Worldwide, 10–20 per cent of drylands are degraded, directly affecting well-being of drylands populations, and indirectly affecting people elsewhere through biophysical and socio-economic impacts. Globally-driven processes, including climate change, have direct impacts on well-being in drylands.

Vulnerability and Human Well-being

There are a number of factors that influence the vulnerability of dryland communities, including:

- biophysical features, especially water availability;
- access to natural and economic resources, levels of development, and conflict and social instability;

- interlinkages between dryland and non-dryland areas through migration, remittances and trade; and

People in the drylands of industrialized countries – such as in Australia and the United States – typically have a diversity of livelihood options, and can adapt more to land degradation and water scarcity more easily than can rural people in drylands in developing countries who directly depend on environmental resources for their livelihoods. They are most vulnerable. Although high land productivity and a strong manufacturing sector, such as in North America, can decrease vulnerability, the distribution of access to natural and economic resources, and participation in decision making trigger the vulnerability pattern.

Desertification is a challenge for development and improving well-being. Globally, some 60 000 square kilometres of productive land and about US$42 billion in income are lost annually, due to declining agricultural productivity. Since 1975, the incidence of drought has increased fourfold from 12 to 48 episodes. Where there is high agricultural dependency, droughts may undercut food security and economic performance, lessening the opportunity to meet MDG 1. In Pakistan, for example, drylands are increasingly threatened by declining soil fertility and flash floods – early warnings of a looming crisis. The seemingly low production potential of drylands has made them less favoured for the systematic investments (in water and land) needed to offset negative effects of land use and sustain their productive capacity. Freshwater availability in drylands is projected to be further reduced from an average of 1 300 cubic metres/person/year in 2000, which is already below the threshold of 2 000 m^3 required for minimum human well-being and sustainable development. In arid and semi-arid areas, water shortages are predicted to be the most significant constraint for socio-economic development. In some countries, the reduced supply of potable water will mean women and girls will be forced to travel longer distances to collect water.

The high number of transboundary aquifers under stress may, in some instances, add a regional dimension to the risk of tensions related to water scarcity. In some situations, adaptation strategies, such as irrigation of water-intensive crops, lead to clashes between rural and urban users, as well as between agriculturalists and pastoralists. In the US southwest for example, multistakeholder dispute resolution mechanisms, including judicial systems and significant technological and financial resources, keep most of these conflicts from turning violent. In areas with higher vulnerability, such as the Sahel, shortages of arable land and water, particularly in drought periods, have sometimes led to violent conflicts along a number of lines of division: rural-urban, pastoralist-agriculturalist and ethnic group-ethnic group (Kahl 2006, Lind and Sturman 2002, Huggins and others 2006). Movement of "dryland refugees" to new areas, including cities, has the potential to create local and regional ethnic, social and political conflict. Seasonal and cyclic migrations are important coping strategies for pastoral dryland peoples. Pastoral societies (found in all regions) are critically exposed to ecosystem change, which can increase their vulnerability, affect their capital stocks, hinder coping strategies, decrease the productive performance of livestock and generate tensions with other herder and host farmer communities.

Given the extent of drylands, the roughly 2 billion people they support and the biological diversity they hold, the maintenance and restoration of their ecosystem functions is essential for achieving the CBD 2010 biodiversity targets and the MDGs. The UN Convention to Combat Desertification (UNCCD) provides the overall framework for addressing land degradation. It is complemented by the CBD, UNFCCC, Agenda 21, WSSD and other multilateral agreements.

The UNCCD supports national action to combat desertification and improve opportunities from land management. This includes the development of national (NAP), sub-regional (SRAP) and regional (RAP) action programmes. By 2006, a significant number of countries had developed NAPS, with 34 in Africa, 24 in Asia, 21 in Latin America and the Caribbean, and eight in Europe. The CBD provides for management based on equitable benefit-sharing, which helps to increase local resource-based income. Successful applications in drylands include co-management initiatives for wildlife, and the development of markets for non-timber forest products (NTFPs). Intergovernmental initiatives, including the WSSD, UNCCD and the UNEP-led BSP, that focus on capacity building and transfer of technology to enhance management, production and marketing, offer opportunities for building on these successes.

Early warning systems (EWS) are widely used to improve the ability to respond to environmental pressures. The UNEP/FAO Land Degradation Assessment in Drylands (LADA) systematically observes land degradation to increase understanding of drought and desertification processes and their effects. In addition, national, sub-regional and global EWS enhance capacity to respond to potential food insecurity. In Eastern Africa, for example, the Intergovernmental Authority on Development (IGAD) links conflict monitoring (through its Conflict Early Warning and Response Mechanism) to environmental EWS (through its Drought Monitoring Centre), because drought and other environmental pressures may trigger pastoral conflict.

Effective responses to the multiple and complex drivers of land degradation demands interlinked approaches, adequate funding and sufficient capacity. For example, attempts to reverse water degradation trends are constrained by a number of factors. They include: poverty, slow economic development, deficiencies in the technical, administrative and managerial capacity of water management institutions, weak national and regional legal frameworks, and a lack of international cooperation. Developing systems for managing water scarcity, which deal with rainwater and run-off, and mediating between competing water claims, including environmental claims, has proved difficult. The failure to harness different kinds of knowledge, including traditional farming knowledge, in management and policy means that the full range of options for improving dryland farming is not taken up. Insufficient funding, including for NAPs, and failure to respond to early warnings, are constraints.

Experience shows that financial investments and loans to dryland farmers can produce significant returns, but this approach continues to be underused. Although women play a pivotal role in environmenta and agricultural management, they have limited support. Institutional and governance factors, coupled with insufficient capacity, limit the financial benefits that producers reap from drylands products, such as crops and NTFPs (Marshall and others 2003, Katerere and Mohamed-Katerere 2005). In 2005, UNCCD COP 7 acknowledged that insufficient decentralization and insecure tenure undermine management and reduce opportunities. Potential income is lost to intermediaries: in Namibia, devil's claw *(Harpagophytum* species) producers receive just a fraction of the retail price, ranging from 0.36 per cent, when dealing with intermediaries, to 0.85 per cent when selling directly to exporters.

Global trade regimes, particularly protectionist tariffs and agricultural subsidies in developed country markets, affect income of drylands producers in developing countries. These tariffs and subsidies have, for example, reduced the competitiveness of developing country cotton, even though developing countries are among the lowest-cost producers. Conflict can also be an important factor inhibiting product and market development in drylands.

Addressing these constraints can improve opportunities for increasing well-being. Options include :

- improving tenure, and recognizing the value of traditional knowledge to encourage farmer investments in soil and water conservation, which lead to more profitable agriculture;
- addressing resource-related conflicts through multilevel environmental and development cooperation, including bringing all stakeholders together to negotiate sharing benefits from interdependent resources, such as transboundary water. This helps to build trust through cooperative environmenta management; and
- ensuring more equitable access to global markets, to improve opportunities for agriculture and ivelihood diversification.

Misusing the Global Commons

Another archetype is a pattern of vulnerability resulting from misuse of the global commons, which include the deep oceans and seabed beyond national jurisdiction and the atmosphere. In some contexts biodiversity (where species concerned are found in the globa commons) and Antarctica are also included in the list of global commons, but the focus here is on the oceans and the atmosphere. The misuse of these global commons leads to the exposure of people and the environment to pollution (such as heavy metals and persistent organic pollutants in the Arctic), to resource depletion (such as in fisheries) and to environmental changes (in particular as a result of climate change). Very often those that are extremely vulnerable to the changes resulting from misuse of the commons are not responsible for the misuse itself.

Global Relevance

Resources that cannot be governed under the norma governance framework of national sovereignty are usually referred to as 'global commons.' The globa commons physically envelop the globe and humanity. The oceans have the character of both a common (re)source – for example, providing large amounts of fish – and a common 'sink' – receiving large amounts of pollution from ships, land and the atmosphere. The atmosphere is a decisive (re)source for life on this planet, both because it protects people from the harmful rays of the sun and provides the climate system, and because the oxygen in its lower parts is also the source of the air most organisms need to sustain life. The atmosphere is heavily misused as a sink for pollution from a wide range of human activities.

Vulnerability and human well-being. Marine living resources provide a significant proportion of protein in the human diet. Two-thirds of the total food fish supply is from capture fisheries in marine and inland waters. However, fisheries are declining, formerly abundant species are now rare, food webs are being altered, and coastal ecosystems are being polluted and degraded. In some cases, fisheries have collapsed, and the livelihoods of entire communities have been destroyed. A well-known example is the collapse of much of the Canadian cod fishery. In the early 1980s, the Canadian catches of Atlantic groundfish peaked, and then declined rapidly.

The Mediterranean Sea is currently part of the global commons, since many surrounding countries have not exercised their right to establish 200-mile exclusive economic zones. As a result of overfishing and pollution in the Mediterranean, catches of the high-value bluefin tuna reached a high of 39 000 tonnes in 1994, but had dropped by nearly half that amount by 2002. More recently, after the decline of traditional stocks, such as cod, attention has turned to deep-sea fishing (deeper than about 400 m), where fish are particularly vulnerable to overfishing because of their slow ability to reproduce. Several deep-sea stocks are now heavily exploited, and, in some cases, severely depleted. A very small number of countries land most of the fish catch from the high seas.

Many coastal communities have no capacity to fish in the global commons of the high seas, and are thus deprived of the food and revenue the resource provides. The disruption of small-scale fisheries by high-technology competition often leads to a vicious cycle of fisheries depletion, poverty, and loss of cultural identity. It can also lead to conflict. An example of the impacts on human well-being from air pollution is the long-range transport of persistent organic pollutants (POPs) and heavy metals, which disproportionately affect indigenous people of the Arctic. These same communities are also vulnerable to the adverse impacts of climate change.

Responses

People from more than 190 countries use the global commons, but no global authority exists to enforce a management regime. Agreements built around consensus are often very weak. In some cases, countries do not sign or accede to the agreements, leading to the "free rider" problem.

The wide range of agreements now covering the use of ocean resources beyond national jurisdiction includes the UN Convention on the Law of the Sea (UNCLOS), the UN Agreement on Straddling Fish Stocks and Highly Migratory Fish Stocks, the Convention on Biological Diversity, the International Plan of Action on Illegal, Unreported and Unregulated Fishing, and a range of regional fisheries agreements. However, management responses have been unable to keep pace with the repeated pattern in deep-sea fishing of exploration, discovery, exploitation and depletion. Gaps in the high seas governance regime contribute to the depletion of deepwater fish stocks. There is a strong need for integrated approaches instead of separate regimes for fisheries, aquaculture, marine mammal conservation, shipping, oil and gas, and mining. A multiplicity of sectoral agreements cannot deal with conflicts across sectors, or with cumulative effects.

Over the past decade, multilateral agreements have been adopted to deal with persistent organic pollutants. The global Stockholm Convention on POPs and the regional UNECE/CLRTAP POPs protocol both seek to phase out the production and use of a number of harmful substances. POPs are also subject to strong policy actions under the European Union, the Convention on the Protection of the Marine Environment of the Baltic Sea Area, the Convention for the Protection of the Marine Environment of the North-East Atlantic (OSPAR), and the North American Agreement on Environmental Cooperation (NAAEC). These overlapping international agreements, together with increasing domestic regulations have, in many cases, resulted in declining pollution levels and reduced threats to human health.

There is no global heavy metals (HM) agreement. The HM agreement with the largest geographical coverage is the 1998 UNECE/CLRTAP Heavy Metals Protocol. HMs are also subject to regulations under the European Union, HELCOM, and OSPAR. Mercury is also targeted under the NAAEC. Global efforts to address mercury led to a mercury assessment, and the UNEP Mercury Programme. HM emission reductions measures, such as limiting allowed emissions from major stationary sources and bans on lead in gasoline, have helped to reduce emissions. Despite these actions, environmental levels of some HMs do not seem to be declining, and in some cases, are even increasing, raising concerns for human health.

It has been possible to misuse the oceans and the atmosphere for long periods of time with only slowly emerging visible repercussions. Their volumes are very large, their composition very complex, the lag times between cause-and-effect are long, and their physical "location" can be distant from people. Furthermore, the response capacity of the international community has been predominantly low, with exception of protecting the stratospheric ozone layer. It has been difficult to overcome the challenges, and to manage these global commons as collective resources of humankind, because of the weak institutional architecture at the global level.

Despite these challenges, international treaty regimes to protect global commons signal an unprecedented level of international cooperation, and are giving rise to a number of policy innovations in global environmental governance, such as emissions trading schemes (the Kyoto Protocol) and shared revenues from using resources (UNCLOS). But reducing vulnerability related to the degradation of global commons requires a number of responses beyond international treaties alone. Some of the opportunities that deserve closer attention are:

- integrating governance from the local to the global level by supporting governance measures at all levels, and going beyond providing resources and capacity building for national agencies in charge of implementing global agreements;
- strengthening the voices of vulnerable communities in global processes, helping to bridge different types of knowledge, and to build a culture of responsibility for action;
- institutionalizing longer time horizons and intergenerational equity in research efforts, impact assessments, decision making and law, which is essential for reversing the pattern of misuse of the global commons, and which will need consistent incremental decisions and policies over years and decades to effect change;
- paying attention to mitigation and adaptation to help the communities most vulnerable to degradation of the global commons in ways that are sensitive to their local cultures, for example in the global treaties that until now have their strongest focus on reducing the degradation of the commons; and
- resolving conflicts with stronger multilateral fish stock management.

Securing Energy for Development

This archetype is about vulnerabilities as a consequence of efforts to secure energy for development, particularly in countries that depend on energy imports. The dramatic increase in energy use in the last 150 years has been a key factor in economic and social development. In those countries and sectors of population that do not yet benefit from modern energy, development is hindered and energy security and increasing energy access are therefore high on the nationa agendas. Vital societal functions depend on reliable energy supply. The dominating energy production patterns (centralized production systems, fossil fuel dominance and lack of diversification) have created increased technical and political risks for disrupted supplies as well as a host of negative health and environmental effects.

Since the 1970s, each 1 per cent increase in GDP in industrialized countries has been accompanied by a 0.6 per cent increase in primary energy consumption. A further increase of over 50 per cent in energy use - mostly in developing countries - is expected in 2-3 decades. In 2000, about 1.6 billion people had no access to electricity while 2.4 billion people still relied on traditional uses of biomass, a burden that falls mainly on women. Although there are no MDGs for energy access, the WSSD warned that without access to modern energy supplies, and fundamenta changes in energy use, poverty reduction and sustainable human development would be difficult to realize.

Oil and gas are expected to remain the dominant sources of energy over the next 2-3 decades, if current trends continue. Energy supply security is becoming a problem, due to increasing competition for oil and natural gas among Europe, the United States and the rapidly-growing economies in Asia. Among the factors affecting supply security are :

- oil exports are from a smaller number of countries;
- geo-political tensions;
- uncertainty over when the global resource base for oil and gas may become critical, with mainstream energy analysis suggesting this is unlikely in the next 2-3 decades, while others believe that oil production already is peaking; and
- impacts of extreme weather events on energy production, such as the heat wave in Europe in 2003, and hurricanes in Gulf of Mexico in 2005.

About 90 per cent of the global anthropogenic greenhouse gas emissions are energy related, and dramatic shifts towards low greenhouse gas emitting production and consumption systems are necessary to address climate change problems, especially in developed and rapidly developing countries.

Oil has become increasingly important in total energy consumption of low-income regions. In contrast, in high-income countries, the share of oil in energy use has declined, although absolute consumption of oil still increases. The share of oil that is imported is increasing in both high- and low-income countries, following a decline in the 1970s and 1980s due to the oil crises. Since the early 1970s, oil intensity has almost halved in high-income regions. Although oil intensity is declining in low-income regions, the ratio is significantly higher, indicating that oil price shocks are having a far greater impact on their economies.

Vulnerability and human well-being. Impacts of energy use on human well-being due to air pollution and climate change, as well as the importance of energy for realizing the MDGs.. For energy-importing countries, securing the supply of affordable energy is directly linked with human well-being. There could be a "vulnerability paradox" regarding energy: the less vulnerable a country's energy sector becomes, the greater the impacts could be from energy problems. Since society has become very dependent on energy, there could even be a "double vulnerability paradox." Both the decreased vulnerability of the energy supply as well as the increased dependency on a reliable energy supply contribute to an increasing vulnerability of society to disturbances in the energy supply. For households, energy becomes an issue of concern with increasing energy prices. This especially affects lower-income groups in industrialized and developing countries. For instance, the United Kingdom has had a fuel poverty strategy since 2001 that recognizes that fuel poverty is caused by a combination of low income, lack of energy efficiency measures and unaffordable energy, especially for the elderly.

For developing countries without fossil fuel reserves, the security of supply is an even more pressing problem. Again, this affects the poorer population groups, because transport and food prices are affected most. Rural areas are especially vulnerable, as are small and medium enterprizes that often cannot cope with the volatility of oil prices. Rises in energy prices also result in macro-economic losses, indirectly affecting human well-being.

Countries have pursued a variety of options to improve their energy security, including diversifying energy supply, improving regional energy trade arrangements, reducing dependence on imports by promoting energy efficiency, using domestic sources and alternative options, including renewable energy. In most countries, the buildup of energy infrastructure is extensively regulated by governments. With liberalization in many industrialized and developing countries over the last decade, this situation has changed. The internal market in Europe had two opposing effects with respect to energy security and the environment. It improved the overall efficiency of the energy system, and created a market for more energy-saving technologies. However, it also made investments that require large capital input, or have long payback times. R&D became more short-term oriented, and budgets were reduced and often not aligned with sustainable development objectives.

Public support remains necessary to stimulate new technologies. Many development strategies treat energy only in the context of large-scale infrastructure projects, where energy access issues are usually ignored, and the focus is on electricity, neglecting fuel availability and rural energy development. Out of 80 MDG country reports, only 10 mention energy outside discussions in relation to environmental sustainability (MDG 7). Only one-third of Poverty Reduction Strategy Papers allocate financial resources to national energy priorities. Implementation of sustainable energy systems is hindered by a number of issues, including a finance gap, subsidies biased towards fossil fuels, lack of stakeholder involvement, and regulatory and sector management problems.

Energy has long been considered the exclusive prerogative of national governance, and with the exception of nuclear energy, has lacked both an organizational home and a coherent normative framework in the UN system. This has begun to change in recent years with energy for sustainable development being discussed as a theme by the Commission on Sustainable Development in 2001 and 2006–07. At the World Summit on Sustainable Development (WSSD), energy received high priority in the action plan. Converging agendas seem to be pushing for strengthened global governance of energy through its links to climate change, poverty (especially MDG 1), health and security. Following the WSSD, a number of multistakeholder partnerships were established to implement various elements of the international energy agenda. As a follow-up to the 2005 G8 Gleneagles energy initiative, the World Bank completed in 2006 an investment framework for clean energy and sustainable development. There has also been some efforts to create mechanisms for coordinating energy work, most recently through UN-Energy, an interagency mechanism established to support the implementation of WSSD energy-related decisions.

The policies to move away from oil dependence have had some impacts in industrialized countries. One of the reasons for the limited impact of policies is the long lifespan (40–50 years and longer) of energy infrastructure. This means that technology and investment decisions from decades ago have created a path dependency for today's production and consumption patterns. It also means that the decisions made today will have major impacts for decades to come, and there are few incentives in place for considering the well-being of future generations.

Given the large scope for synergies among policies related to energy security, health and air pollution and climate change, there are many opportunities to reduce vulnerability of people and communities, including:

- focusing energy policies on improving access to appropriate energy services for the most vulnerable, such as women, the elderly and children, as part of broad development planning;
- improving the opportunities for the most vulnerable to have voice in energy issues, for instance, in designing new energy systems;
- investing in the diversification of both centralized and decentralized technologies, with technology transfer playing an important role; and
- strengthening the capacity for sustainable energy technology innovation and production in cooperation with vulnerable communities, as a way to create jobs and increase coping capacity.

Coping with Multiple Threats in Small Islands

Small Island Developing States (SIDS) are vulnerable to climate change impacts in the context of external shocks, isolation and limited resources, creating another archetype of vulnerability. SIDS are highly prone to natural disasters, such as tropical storms and storm surges. Limited institutional, human and technical capacities highly constrain their ability to adapt and respond to climate change, variability and extremes. Current vulnerabilities are further exacerbated by growing populations. For example, the total fertility rate of most Pacific islands is greater than four. The international trading regime and WTO compliance are increasingly demanding for SIDS. With eroding access to protected markets for their export commodities, such as sugar, bananas and tuna, and with declines in commodity prices triggering economic volatility, they are highly sensitive to globalization and trade liberalization.

SIDS are located in the Pacific, Indian and Atlantic Oceans, and the Wider Caribbean and South China Seas. In UNEP regional terms, 6 SIDS are in Africa, 23 in Latin America and the Caribbean, and 22 in Asia and the Pacific. The Environmental Vulnerability Index (EVI) scores for 47 SIDS illustrate that none are ranked resilient and almost three-quarters are highly (36 per cent) or extremely (36 per cent) vulnerable. EVI was prepared by various organizations, including UNEP.

Vulnerability and human well-being. Natural hazards have severe adverse impacts on lives and socio-economic development in SIDS. A high proportion of the total population of 56 million is frequently exposed to natural hazards. For example, in 2001 nearly 6 million people were affected by natural disasters in the Caribbean.

Sea-level rise, and the increasing frequency and severity of extreme events threaten livelihoods and limit adaptation options. These pressures have forced some people to abandon their homes and assets, and to migrate to other countries. New Zealand, for example, amended its Government Residence Policy in March 2006 to allow a small number of citizens from Tonga, Tuvalu, Kiribati and Fiji to immigrate each year. Sea-level rise is likely to induce large-scale migration in the longer term, and large migrations have at times led to conflict. Abandoning islands would also result in the loss of sovereignty, and highlights the need to reconsider traditional development issues as matters of national and regional security, as well as issues of equity and human rights.

Climate-related hazards cause socially differentiated impacts, and tend to affect the poor and disadvantaged groups disproportionately. Most exposed to hazards are people living on atolls and low-lying islands, and in high-risk coastal settlements with substandard housing and infrastructure. The livelihoods most affected include those depending on climate-sensitive natural resources, such as subsistence and commercial farming, and on coastal tourism. The most severe impacts on human well-being include the loss of livelihood assets, displacement, increased water- and vector-borne diseases, and the loss of life in natural disasters. The loss of livelihood assets is predominantly caused by reduced or lost ecosystem services due to recurring natural hazard impacts, loss of productive land due to coastal erosion, salinization of land and irrigation water, estuaries and freshwater systems, and other forms of environmental degradation, such as deforestation. In addition, degradation and overexploitation harm resources such as coral reefs, seagrass beds and mangroves that provide a natural coastal protection as well as the basis for subsistence and commercial activities. Hoegh-Guldberg and others estimate that coral bleaching will reduce future GDP by 40–50 per cent by 2020 in smaller Pacific islands. Furthermore, SIDS are faced with biodiversity loss and impacts on agriculture, due to invasive alien species.

Deteriorating resource access has led to growing competition at community, national and regional levels, though pressures are spatially variable. Further stresses, including social pressures from eroding customary resource tenure and security of land titles, have been highlighted as key issues for the management of some marine ecosystems.

Higher exposure to natural hazards can have negative impacts on tourist infrastructure and investments, and can reduce tourism income. At the same time, tourism adds to increased pressures on ecosystems. In some coastal locations, inappropriate development in risk-prone areas, due to inadequate consideration of impacts of natural hazards and climate change effects, demonstrates a failure to adapt.

Responses

Recognizing the vulnerabilities of SIDS, the Barbados Programme of Action for the Sustainable Development of Small Island Developing States was adopted internationally in 1994. The Commission on Sustainable Development reviewed the implementation of the Barbados Programme of Action in 1996 and 1998. In 2005, the programme was reviewed at a UN Conference in Mauritius, at which the opening statement pointed out that a decline of international support and resources had hindered implementation. The Mauritius Strategy was adopted at the 2005 conference, laying out a comprehensive multilateral agenda for the sustainable development of SIDS.

The Cooperative Initiative on Invasive Alien Species (IAS) on Islands deals with invasive species that threaten biodiversity, as well as agriculture and human well-being. Innovative initiatives also link ecotourism with eradication of IAS.

While some adaptation options are already being implemented in SIDS, specific adaptation strategies offer opportunities for more efficient adaptation, including the use of traditional knowledge based on typical regional or cultural conditions. For example, traditional food preservation techniques, such as burying and smoking food for use in drought periods, can improve food security in rural areas. Traditional building materials and designs help reduce infrastructure damage and loss from natural hazards. Renewable resources, such as biofuels (such as bagasse), wind and solar power, show a great potential for energy diversification, and for improving the energy resource potential and energy supply for SIDS. This can also increase resilience in the face of recurring extreme events.

To achieve this overarching goal of successfully improving human well-being in SIDS, vulnerability and adaptation assessments need to be further mainstreamed into national policies and development activities at all levels and scales. A number of options are available to reduce vulnerability, and to build capacities in SIDS:

- enhancing early warning systems to support disaster preparedness and risk management systems helps adaptation to short-term variability (Yokohama Strategy and Plan of Action for a Safer World 1994 and the Hyogo framework) ;
- improving integrated planning for climate-robust, long-term development, especially that of ivelihood assets, improves access to resources for local people. Water resource and Integrated Coastal Zone Management (ICZM) can contribute to improving the long-term adaptive capacity of vulnerable communities. This requires governance systems that take possible long-term changes into account;
- using participatory approaches to integrate traditional ecological knowledge in conservation and resource management empowers communities for disaster preparedness and resource management;
- developing technologies for reducing vulnerability can shift from a "controlling nature" to a "working with nature" paradigm. This includes the technology and capacity to assess impacts and adaptation options, document traditional coping mechanisms and develop alternative energy solutions;
- investing in improved regional cooperation can better address environmental challenges and improve coping capacity. An example

would be development and strengthening of globa and regional bodies, such as Alliance of Smal Island States (AOSIS) and the Indian Ocean Commission, to build early warning systems for environmental stresses;

- strengthening of cooperation and partnerships at the national, regional and internationa evels, including pooling of resources for the implementation of activities and Multilatera Environmental Agreements (MEAs); and
- recognizing in international negotiations that basic rights laid down in the Universa Declaration of Human Rights are at risk in the case of climate change effects on atoll countries.

Technology-centred Approaches to Water Problems

Poorly planned or managed large-scale water projects that commonly involve massive reshaping of the natural environment can create another archetype of vulnerability. Examples include certain irrigation and drainage schemes, the canalization and diversion of rivers, large desalinization plants and dams. Dam projects are prominent and important examples, although many of the conclusions often apply to other vulnerability-inducing water management schemes. Dams have both positive and negative impacts: they satisfy human needs (water for food security and renewable energy), and protect existing resources by providing flood control. However, they may have severe impacts on the environment through river fragmentation, and on social structure. Some dams provide benefits without major negative effects. But many do not due to the inadequate consideration given to social and ecological impacts from poor dam planning and management. This is a result of the prevailing technology-centred development paradigm. Reducing vulnerability here means either to reduce the negative consequences of these projects, or to find alternative means to fulfil the demand for energy, water and flood protection.

The dynamics described here occur worldwide. Important examples are the planned Ebro water scheme in Spain, large-scale water management schemes in the US southwest, the Narmada in India, the Nile in Africa and the Three Gorges Dam in China. Major irrigation schemes built in the 20th century and new, multifunctional mega-dams (over 60 m in height) have had significant impacts on water resources. There are more than 45 000 large dams in 140 countries, about two-thirds of these in the developing world. The actual trend is characterized by a decline in the annual number of new large dams, while no decline is observed for the mega-dams. The geographical distribution of new dam construction continues to shift from the industrialized countries to the newly industrialized and developing countries. The effects of these large-scale installations are rarely confined to the local area, but can assume far-reaching and even international proportions.

Currently, large dams are typically built in remote areas of developing countries. The integration of such peripheral regions into the world market through dam projects leads to an extensive transformation of social conditions for the indigenous population. Consideration must be given to the social consequences, which may range from resettlement of the local population, to intensification of economic disparities, and domestic and international conflicts (McCully 1996, Pearce 1992, Goldsmith and Hildyard 1984). According to estimates, 40–80 million people have been forced to leave their homes since 1950 because of large dam projects. Forced resettlement, lack of stakeholder participation in planning and decision making, and lack of sharing in the benefits of the projects may marginalize and victimize the local people in development. The distribution of the benefits gained from dam construction (power generation and irrigated agriculture) can be very uneven, reinforcing the widening of social and economic disparities and poverty.

Tensions may build up, and can escalate into national and international conflicts. Although widespread organized violence is rare, local protests against large water projects are common. Despite high levels of political attention to future "water wars" between states, cooperation between states has been more common than conflict over the last half of the 20th century. A comprehensive analysis of bilateral and multilateral state-to-state interactions over water between 1948 and 1999 found that of more than 1 830 events, 28 per cent were conflicts, 67 per cent were cooperative, and the remaining 5 per cent were neutral or not significant. International water cooperation institutions, such as basin commissions, have fostered international cooperation, for example in the cases involving the Itaipu and Corpus Christi dams in Argentina, Brazil and Paraguay. In some cases, a key to fostering cooperation appears to be moving parties, often through external facilitation, from asserting competing rights to water to identifying needs for water, and finally to negotiating the sharing of the benefits of water. Further examples of cooperation involve the Zambezi, Niger, Nile and Rhine rivers.Other negative impacts on human well-being are health hazards in the form of water-based vectors (for example, mosquitoes and snails), which occur due to the changes in the run-off regimes. This exacerbates the risk of malaria and other diseases in many subtropical and tropical regions.

Responses

In 2000, the international multi-stakeholder World Commission on Dams (WCD) evaluated the development effectiveness of large dams, and developed international guidelines for dam building.

Building synergies between biodiversity concerns (as in the CBD, RAMSAR Convention on Wetlands and the Convention on Migratory Species), and development is an important concern. As a follow-up to the WCD framework, the UNEP Dams and Development Project (UNEP-DDP) was launched in 2001. Recognizing that for many developing countries hydropower and irrigation remain priorities to meet energy and food security needs, UNEP-DDP focuses on how to support building and management of dams sustainably. At national and sub-regional levels, countries have responded by increasingly accepting social and environmental impact assessments (EIAs) of large dam projects prior to construction. The trend towards shared river management, acknowledged in the 1997 UN Convention on the Non-Navigational Uses of International Watercourses, has created new opportunities for addressing such concerns.

Nevertheless, the effectiveness of these measures is mixed. In some places, it is evident that stakeholder expectations regarding participation, transparency and accountability in dam planning and development is changing. The WCD recommendations provided a new, authoritative reference point for NGOs trying to influence government decisions, but has had different levels of success. The value of cooperation between states is increasingly recognized, but in practice this has played out in different ways. For example, the controversial Ilisu Dam project in Turkey came to a halt in 2001 when the European construction firms withdrew from the project, citing outstanding economic and social issues, and the difficulty of meeting conditions imposed on their effort to procure a US$200 million export credit guarantee from the British government. In contrast, the World Bank and the African Development Bank forged ahead with the controversial US$520 million Bujagali Dam project in Uganda despite strong transnational NGO opposition, and the earlier withdrawal from the project by bilateral funding agencies in the United Kingdom, France, Germany, Sweden and the United States.

Several relevant international policy initiatives deal with problems of insufficiently and inequitably fulfilled water demand. One important aspect of MDG 7 on ensuring environmental sustainability is to "reduce by half the proportion of people without sustainable access to safe drinking water." The implementation plan calls for an approach to "promote affordable and socially and culturally acceptable technologies and practices." These needs can be met, as suggested by the World Water Vision, through a mix of large and small dams, groundwater recharge, traditional, small-scale water storage techniques and rainwater harvesting, as well as water storage in wetlands.

It is clear that maladapted and mainly supply-oriented technological approaches will, at least in the medium-term, fail to realize the desired development benefits.

Well-planned water management can reduce vulnerability, and contribute to development. There are a number of options :

- improving access to water as an essential asset for household needs and agricultural production. Distributional aspects should be given much more attention;
- increasing opportunities for more effective local participation in basin and catchment management, as local rights and values may be in conflict with those held by the state. This requires supportive and inclusive institutions, and governance processes;
- trading, including the import of "virtual water" via food imports, may substitute for irrigation water consumption in arid regions;
- improving cooperative water basin management can increase development opportunities, and reduce potential for conflict. Developing transboundary river basin institutions offers important opportunities for building on environmental interdependence to foster collaboration and contribute to conflict prevention. The SADC Water Protocol of 2000, the Nile Basin Initiative (NBI) and the Niger Basin Authority (NBA) are good examples in Africa of riparian dwellers and stakeholders developing shared visions for water and development, while integrating international legal norms, such as prior notification and causing no significant harm; and

- investing in local capacities and employing alternative technologies can improve water access and use. This strategy is an important way to enhance coping capacities, and ensure consideration of a broader range of alternatives to conventional, large-scale solutions.

Rapidly Urbanizing the Coastal Fringe

Rapid and poorly planned urbanization in often ecologically sensitive coastal areas increases vulnerabilities to coastal hazards and climate change impacts. In recent decades, many of the world's coastal areas have experienced significant and sometimes extremely rapid socio-economic and environmental changes. Limited institutional, human and technical capacities have led to severe hazard impacts, and constrain the ability of many coasta communities, particularly those in the developing world, to adapt to changing conditions.

Many of the world's coastal areas have been experiencing rapidly-growing concentrations of people and socio-economic activities. The average population density in coastal areas is now twice as high as the global average. Worldwide, more than 100 million people live in areas no more than 1 m above sea level. Of the world's 33 mega-cities, 26 are located in developing countries, and 21 are in coastal areas.

Much of this development has been occurring in low-lying floodplains, river deltas and estuaries that are highly exposed to coastal hazards, such as storms, hurricanes, tidal surges, tsunamis and floods. In many cities, major rezoning of former industrial waterfront areas is being undertaken in flood-prone locations to accommodate the tremendous requirement for housing. Examples include Brooklyn and Queens in New York, and the Thames Gateway, a 60-km-long corridor along the Thames River between London and the Thames Estuary that is currently undergoing considerable urban regeneration.

Poor urban planning and inappropriate development in highly exposed coastal locations, in combination with rapid population growth, sea-level rise and other climate change impacts, have led to a considerable increase in socio-economic impacts from coastal hazards. The EM-DAT global datasets on extreme events indicate that annual economic losses from extreme events have increased tenfold from the 1950s to the 1990s. In the decade between 1992 and 2001, floods were the most frequent natural disaster, killing nearly 100 000 people and affecting more than 1.2 billion people. Munich Re documented an increasing concentration of the loss potential from natural hazards in mega-cities. Only a small proportion of these losses were insured.

Environmental change is expected to exacerbate the exposure of many coastal urban areas to natural hazards from rising sea levels, increased erosion and salinity, and the degradation of wetlands and coastal lowlands. There is also a concern that climate change might, in some areas, increase the intensity and frequency of coastal storms and hurricanes, but there is no scientific consensus. In a recent global assessment of storm surges, Nicholls estimated that in 1990 some 200 million people were living in areas vulnerable to storm surge flooding. The North Sea, the Bay of Bengal and East Asia are considered as notable hot spots, but other regions, such as the Caribbean, and parts of North America, Eastern Africa, Southeast Asia and Pacific states are also vulnerable to storm surges.

Increasing development in coastal areas causes fragmentation of coastal ecosystems and conversion to other uses, including infrastructure and aquaculture development, and rice and salt production. This negatively affects the condition and functioning of ecosystems, and their ability to provide ecosystem services. An assessment of the status of the world's mangroves FAO found that their extent has been reduced by 25 per cent since 1980.

The relationship between increasing urbanization and growing vulnerability to natural hazards is most pronounced, but not exclusive in developing countries due, to a large extent, to rural-urban migration. This often affects cities facing severe constraints on their institutional, human, financial and technical capacities to develop integrated approaches to urban planning. As a result of the lack of affordable housing options, poor migrants tend to inhabit informal settlements, which are often located in the most undesirable and hazardous areas of the city. According to UN Habitat, more than 750 million of the world's more than 1 billion poor people live in urban areas, without adequate shelter and basic services. Unsafe living conditions, the lack of secure livelihoods and access to resources and social networks, and exclusion from decision making processes limit the capacity of poor urban people to cope with a range of hazards.

Estimates of the number of additional people at risk from coastal flooding in the future vary widely, but all indicate a considerable increase. For example, Nicholls estimates that the number of people living in areas vulnerable to storm surge flooding to increase by nearly 50 per cent (or 290 million) by the 2020s compared to 1990, while Parry and others estimate about 30 million more people at risk from coastal flooding due to climate change by the 2050s and 85 million more by the 2080s.

Response

In recent decades, particularly since the 1990s, the dramatic increase of losses and suffering due to natural disasters has brought the issue of disaster risk reduction increasingly onto the political agenda. From the International Decade for Natural Disaster Reduction (IDNDR) to the International Strategy for Disaster Reduction (ISDR), to the Hyogo Framework of Action, the disaster risk reduction community has been calling for renewed commitment, and the integration of disaster risk reduction, with the pursuit of sustainable development as a strategic goal.

The Hyogo Framework calls for the incorporation of disaster risk assessments into the urban planning and management of disaster-prone human settlements. It prioritizes the issues of informal or non-permanent housing, and the location of housing in high-risk areas. This reflects the ISDR estimate that 60–70 per cent of urbanization in the 1990s was unplanned. One consequence of this framework is that international organizations, like UNESCO, reviewed their present activities with respect to the suggested actions for disaster risk reduction. Most of the urbanization challenges are still the result of a lack of integrated environmental and urban planning. Policies for more sustainable patterns of urbanization are frequently not implemented. Short-sighted concessions for economic gain, weak institutions and corruption are major factors in the proliferation of planning "oversights," "exceptions" and other forms of inappropriate development in urban areas.

Networks such as the African Urban Risk Analysis Network (AURAN) aim at mainstreaming disaster risk reduction in the management of urban planning and governance in Africa. Here, community-based action research is supported. Projects like "Engaging in awareness-raising activities and household surveys on local perceptions of flood risk in flood-prone districts of Saint Louis, Senegal" reduce vulnerabilities locally, and generate transferable knowledge for other cases.

Science increasingly recognizes the importance of sustainable resource management and biodiversity for ecological resilience and livelihood security in the face of extreme environmental shocks. For example, the detrimental consequences of the loss of coastal ecosystems and their buffering capacity against natural hazards has recently been documented in relation to the 2004 Indian Ocean Tsunami and the impacts of Hurricane Katrina on New Orleans.

Environmental actions that reduce vulnerability are seldom promoted in disaster reduction strategies, and many opportunities to protect the environment and reduce disaster risk are missed. Integrated coastal zone management (ICZM) and even further integrated coastal area and river basin management (ICARM) are important tools in reconciling multiple uses of coastal resources and promoting ecological resilience. They provide an institutional framework to implement, enforce, monitor and evaluate policies for the protection and restoration of coastal ecosystems, and to place more value on the goods-and-services (cultural values, natural protection of coastal zones, recreation and tourism and fisheries) they provide. There are significant opportunities to reduce hazard vulnerability:

— integrating of risk reduction and adaptation strategies with existing sectoral development policies in areas such as integrated coastal zone management, urban planning, health care planning, poverty reduction, environmental impact assessment and natural resource management;
— strengthening education and awareness raising to deal with the multiple risks associated with rapid coastal urbanization, and with possible response options;
— providing more opportunities for local participation in urban development. The challenge for institutional development is to be responsive to change. One approach is to focus on processes in which local users become active "makers and shapers" of the rights, management and use regimes upon which their livelihoods are based. Participation of women is a critical component in such approaches ; and
— green engineering can help to protect coastlines using mangroves and reefs. It can help to maintain forests and protect soils to avoid the risk posed by landslides, floods, drought and tsunamis.

Challenges Posed by Patterns of Vulnerability

The seven patterns of vulnerability show how environmental and non-environmental changes affect human well-being. Some of the different human-environment systems throughout the world share certain vulnerability-creating conditions. The different patterns reflect vulnerabilities across the full range of geographic and economic contexts: developing and industrialized countries, and countries with economies in transition. This allows putting particular situations within a broader context, providing regional perspectives, and showing important connections between regions and globally, as well as possible opportunities to address the challenges in a more strategic manner. Furthermore, the analysis of the archetypes underlines findings of other vulnerability research:

- Research on the underlying causal structures of human vulnerability to environmental change increasingly recognizes that vulnerability arises through complex interactions of multiple sociopolitical, ecological and geophysical processes that operate in different areas and at different times, resulting in highly differentiated impacts in and across regions, social groups and individuals.
- Environmental risks affect a wide range of natural, economic, political and social activities and processes. Therefore, vulnerability reduction should be integrated as a strategic goal into overall development planning across many sectors, including education, health, economic development and governance. Reducing vulnerability in one area often results in increasing vulnerabilities far away, or moving it into the future, which needs to be taken into account.
- Environmental change has the potential to spur conflict. However, managed environmental change (for example conservation and cooperation) can also make tangible contributions to conflict prevention, de-escalation and post-conflict reconstruction.
- Human vulnerability and livelihood security are closely linked to biodiversity and ecosystem resilience. Sustainable environmental and resource management is important in poverty and vulnerability reduction. Extreme events, such as the Indian Ocean Tsunami, show that environmenta degradation and poorly planned development activities increase communities' vulnerability to shocks.
- Vulnerability is determined, to a large extent, by a lack of options, due to the unequal distribution of power and resources in society, including the most vulnerable population groups throughout the world, such as indigenous people, and the urban or rural poor. Economic sectors

heavily dependent on environmental services are also vulnerable. Resilience increases with diversification of livelihood strategies, and with access to social support networks and other resources.

- For successful use of vulnerability research findings, the policy arena should recognize that vulnerability arises from multiple stressors, which are dynamic over space and time. If vulnerability is reduced to a static indicator, the richness and complexity of the processes that create and maintain vulnerability over time are lost.
- The analysis of the patterns of vulnerability also helps identify a number of opportunities to reduce vulnerability and improve human well-being. Taking these opportunities would also support reaching the MDGs and examples of this are given in which also illustrates how vulnerability works against the achievement of the goals.

Opportunities for Reducing Vulnerability

Policy-makers can use vulnerability analysis to target policies for groups that most need them. Vulnerability analysis helps to examine the sensitivity of a human-environment system (such as a watershed or coastal town) to various social and environmental changes, and its ability to adapt or accommodate such changes. Therefore, evaluations of vulnerability include attention to exposure, sensitivity and resilience to multiple pressures. The evaluations consider the degree to which a system is affected by particular pressures (exposure), the degree to which a set of pressures affect the system (sensitivity), and the ability of the system to resist or recover from the damage. Policies can address each of these components of vulnerability. The analysis, which is most often at the sub-national level is, however, frequently hampered by lack of and/or unreliable data, as well as the challenges of showing the links between environmenta degradation and human well-being.

The archetypes of vulnerability described above highlight responses that have been taken, primarily at the global or regional level, to address patterns of vulnerability. They also point to opportunities for addressing vulnerability by reducing exposure and sensitivity, and through enhancing adaptive capacity. Many of these opportunities are not directly related to environmental policy processes but to poverty reduction, health, trade, science and technology, as well as to general governance for sustainable development. Given the localized nature of vulnerability to multiple stresses, opportunities exist for national decision-makers to target the most vulnerable groups. Decision-makers should clearly identify provisions in their own policies that create and reinforce vulnerability in their countries, and deal with them. At the same time, collaboration at regional and international levels plays a supportive and important role. The opportunities underline the importance of increasing awareness worldwide about the consequences of policy choices for people and the environment in other countries.

Integrating governance across levels and sectors Increasing the coping and adaptive capacity of the most vulnerable people and communities requires integration of policies across governance levels and sectors, and over time to address the coping and adaptive capacities of future generations. A consistent focus on increasing the well-being of the most vulnerable can involve costs for other actors, but it helps promote equity and justice. For some issues, there are clearly win-win situations between short- and long-term goals and priorities, but for many there are considerable trade-offs, not necessarily on a societal level, but clearly for certain groups or sectors in society, and even for individuals. Opportunities include the integration of knowledge and values to underpin and support institutional design and compliance. This involves integrating local and global knowledge, for example on impact and adaptation, and integrating concern for neighbours with concern for all humanity and future generations.

Strengthening coping and adaptive capacity, and reducing the export of vulnerability require much more cooperation among different governance levels and sectors. Such integrated governance requires mutually supportive policies and institutions at all governance levels, from the local to the global. This can be a considerable challenge, as illustrated repeatedly in the implementation of MEAs. In many cases, it requires higher governance levels to provide the resources, knowledge and capacity at lower levels to implement plans and policies. This is in line with the Bali Strategic Plan on Technology Support and Capacity Building (BSP), as well as other capacity-building initiatives. For example, adaptation to climate change among vulnerable communities in the Arctic needs support from national governments and regional organizations. To facilitate successful adaptation, stakeholders must promote and enable adaptive measures. In addition, governments should consider revising policies that hinder adaptation. Self-determination and self-government, through ownership and management of land and natural resources, are important for empowering indigenous Arctic peoples to maintain their self-reliance, and to face climate change on their own terms. Another related strategy of integrating governance across levels is special organizational forms that facilitate cross-level interaction, such as co-management of natural resources.

Different sectoral priorities should be reconciled and integrated through cooperation and partnership, especially when there are trade-offs between them and these affect vulnerability. One strategy is to integrate, in organizational terms, a focus on strengthening coping capacity and reducing export of vulnerability. For example, when councils, task forces, even ministries are set up, their mandates should cover inter-related sectors, and their staff should have the appropriate training and attitudes to implement broader mandates. Another strategy has been to "mainstream" attention to vulnerability through policy. Mainstreaming of the environment has been tried at various governance levels, including in the UN system, with varying degrees of success. A third strategy is to ensure that planning and governance processes include all relevant stakeholders from various sectors, as in successful integrated coastal zone management. A fourth strategy is to address integration between environment and other sectors, using economic valuation, which raises the parity of natural capital in comparison with other types of capital.

The integration of longer time horizons in governance is an even larger challenge given that decision making in governments and other sectors of society tends to be biased towards much shorter time horizons than sustainable development and the well-being of future generations require. Strategies that change the time horizon of decision-makers should be further explored. Such strategies can include: setting clear long-term goals and intermittent targets, extending the time horizons considered in formal planning, developing indicators and accounting measures that illustrate intergenerational impacts, and the institutionalization of long-term liability from harmful activities. These strategies are unlikely to be implemented, however, unless people across societies expand their time horizons for development.

Improving Health

The well-being of present and future generations is threatened by environmental change and social problems, such as poverty and inequity, which are contributing to environmental degradation. Preventive or proactive solutions for many contemporary health problems need to address the links among environment, health and other factors that determine well-being. Opportunities include better integration of environment and health strategies, economic valuation, targeting the most vulnerable, education and awareness, and the integration of environment and health into economic and development sectors.

Measures to ensure ecological sustainability to safeguard ecosystem services will benefit health, so these are important in the long-term. The emphasis on environmental factors has been a central part of the public health tradition. In recent years, several international policies made provisions for improved consideration of health in development. Global initiatives include the World Health Organization's 2005 recommendations for health impact assessment. At the regional level, the Strategic Environmental Assessment Protocol to the UNECE Convention on Environmental Impact Assessment emphasizes consideration of human health. More effective impact assessment procedures are needed in both developed and developing countries. Economic valuation can help ensure that environment and health impacts are given adequate consideration in policy. An integrated economic analysis of such impacts can capture the hidden costs and benefits of policy options, as well as the synergies and institutional economies of scale that may be achieved through complementary policies that support sustainable development.

In most countries, mainstreaming of environment and health considerations into all government sectors and economic endeavours remains a challenge. Policies and practices regarding health, environment, infrastructure and economic development should be considered in an integrated manner. As environmental pollutants affect health through a variety of pathways, environmental monitoring and epidemiological surveillance systems should be strengthened. Health indicators and strategies are needed for specific groups at risk, such as women and children, the elderly, the disabled and the poor. It is important to raise awareness not only in the health sector, but also in sectors such as energy, transport, land-use development, industry and agriculture, through information on the likely health consequences of decisions. Not only health professionals, but also all other stakeholders need the means to evaluate and influence policies that have impacts on health. A better understanding of the dynamic linkages between ecosystems and public health is leading to new and diverse opportunities for interventions early in processes that could become direct threats to public health. Building awareness about environment and health problems, tools and policy options requires sustained and comprehensive communication strategies.

Resolving conflict through environmental cooperation Despite the decrease in civil wars globally in recent years, millions of people continue to be displaced and negatively affected by violent conflict. Armed conflict often, but not always, causes heavy damage to the environment. It reduces societal capacity to adapt to global change, while making sustained environmental management difficult. Reducing violent conflict, whether related to natural resources or not, would reduce a major source of vulnerability, and would better support human well-being in many parts of the world. Environmental cooperation offers several opportunities for achieving these ends.

Policy tools aimed at identifying the contribution of the environment to violent conflict and breaking those links would help redress key stresses. Developing and deploying such tools requires collaboration across a range of environment, development, economic, and foreign policy institutions, including the UN agencies. Such collaboration recognizes interlinkages across the biophysical components of the environment as well governance regimes. Environmental assessment and early warning activities by UNEP and other stakeholders can play an active role in collecting and disseminating lessons learned. This may support the implementation of the UN Secretary-General's call during the UN General Assembly in 2006 for integrating environmental considerations in conflict prevention strategies.

Environmental cooperation has historically had two main areas of focus. At the international level, emphasis has been on multilateral treaties aimed at mitigating the effects of global change. At the sub-regional level, cooperation has focused on equitable sharing of natural resources, such as regional seas and shared water resources, as well as on improving conservation, through transboundary conservation areas (also known as transfrontier parks), to support integration and development-related activities such as tourism. Environmental cooperation – for conflict avoidance and peacemaking – could be employed across all levels of political organization.

In the rush to pursue policy interventions to sever the links between environment and conflict, analysts and practitioners alike have neglected the prospect for building upon environmental interdependences to achieve confidence building, cooperation and, perhaps, peace. Environmental peacemaking is a strategy for using environmental cooperation to reduce tensions by building trust and confidence between parties in disputes. Environmental peacemaking opportunities will remain untested and underdeveloped until more systematic policy attempts are made to achieve these windfalls in a larger number of cases across resource types and across political levels.

Pursuing environmental pathways to confidence building would capitalize on environmenta interdependence, and the need for long-term, iterated environmental cooperation to reduce conflict-induced vulnerability and improve human-well-being. Such policy interventions could:

- help prevent conflict among states and parties;
- provide an environmental lifeline for dialogue during times of conflict;
- help end conflicts with environmental dimensions; and
- help restart economic, agricultural and environmental activities in post-conflict settings.

Not all environmental cooperation lowers vulnerability and increases equity. Systematic assessment of experiences can increase opportunities. Comparing lessons learned across environmental peacemaking cases helps identify environmental management approaches that instigate, rather than ameliorate conflict, such as early examples of transboundary peace parks that neglected wide consultation with local peoples. The ultimate goal of pursuing environmental peacemaking opportunities is reduction of vulnerability and assaults on human well-being created by the still-numerous local, national, and regional conflicts.

Pursuing environmental peacemaking opportunities will require focusing on local, national and regiona institutional settings, rather than the historical emphasis on multilateral environmental agreements. Trying to capture these environmental and conflict prevention benefits requires considerable capacity building among stakeholders, including public and private interests in the conflict, as well as facilitators, such as bilateral donors or UN entities.

Strengthening Local Rights

The fast-paced changes of social and political values create challenges for developing effective responses that address human vulnerability and well-being, while ensuring complementarity among priorities. Strengthening local rights can offer opportunities for ensuring that local and national conservation and development priorities are recognized at higher levels of decision making.

Globalization has resulted in a growing emphasis on free exchange of commodities and ideas, and individual ownership and rights. In some circumstances, this may not support national or regional development goals. Changing values associated with gender, traditional institutions, and democracy and accountability make the management of environmental resources extremely complex, and present challenges for institutional development. For example, the authority and right of both the state and traditional institutions to manage are increasingly contested. This is evident in conflicts around conservation areas, water and forests. Such conflicts often have negative consequences for conservation and livelihoods, and may also have regional implications, where resources are shared.

Mediating these different interests and perspectives requires responses at the national, regional and global levels. Developing more inclusive institutions that recognize the rights and values of local natural resource users can be an effective response, and can facilitate the inclusion of local concerns into governance processes at a higher level. This can also lead to better information sharing, and more equitable distribution of financial and other resources. Inclusive processes can reduce the tension between local values and rights, and those held by state institutions. Making these approaches effective requires investing in capacity building. Scaling these approaches up to the national or regional level can be appropriate, particularly where resource use has implications for users elsewhere, as in the case of water management. Recognizing existing local institutions, including common property institutions, instead of creating new institutions, may be environmentally and socially beneficial, especially where they have a high degree of local legitimacy.

Building better links between local aspirations and the strategies and policies adopted at the global level is more challenging. It is constrained by international law and governance, but is not impossible. Building negotiating capacity can be an important strategy for increasing the development focus of international governance systems. In some sectors, regional cooperation has proved effective in creating synergies between global governance and development objectives. Promoting freer and fairer trade Trade has far-reaching effects on livelihoods, well-being and conservation. Freer and fairer trade can be a useful tool for promoting growth and reducing poverty, improving resilience through transfer of food and technology, and improving governance.

Environmental and equity issues should be at the centre of global trade systems. This is particularly important to ensure that poor people are not taken advantage of when it comes to trade, especially in products, such as hazardous materials that threaten well-being. The trade regime, particularly in agriculture and textiles, is characterized largely by preferential trade agreements (PTAs), bilateral agreements and quotas. High-income countries negotiate bilateral PTAs with poor countries, but such agreements cause more harm than good.

Poor countries, which have abundant labour, are expected to gain from access to larger markets elsewhere, and high-income countries should ensure them such market access. Since small countries have smaller internal markets, lowering trade barriers would provide them with opportunities to exploit economies of scale, so that the poor can garner employment and better wages. Most models show the liberalization of trade under the current Doha round in the World Trade Organization is expected to reduce poverty, particularly if developing countries adjust their policies accordingly. Trade facilitates learning-by-doing, which can drive higher productivity and industrialization. Contact between industrialized and developing countries can be an effective vehicle for diffusion of best practices, through the transfer of capital and knowledge. As poor countries, particularly primary commodity exporters, are vulnerable to price shocks and other market failures, diversification is a good option for reducing vulnerability, and may contribute to sustainable natural resource use.

Higher levels of income, sophisticated markets and the increased power of non-state actors may enhance the prospects for democracy and liberty. Since trade requires large amounts of arm's-length transactions, better institutions are required for it to work smoothly. Trade may not only raise incomes, but also indirectly and directly promotes better international governance, societal welfare, and international and civil peace, which reinforces and is reinforced by prosperity. Trade, like almost all other economic activity, creates winners and losers, and carries externalities. For some, adjustment costs of increased competition can be high. These problems could be addressed by compensating losers, and encouraging increased adaptation through better public investment in education and infrastructure. Trade contributes most to increasing incomes when combined with good governance. Good governance, local capacities to regulate trade, and the regulation of industry in ways that encourage the adoption of best practices all help mitigate externalities, including those stemming from disposal of hazardous waste and pollution from increased consumption.

Securing Access to and Maintaining Natural Resource Assets

For many people in developing countries as well as indigenous peoples, farmers and fishers in developed countries, secure entitlements to productive assets, such as land and water, are central to ensuring sustainable livelihoods. Continued natural resource availability and quality, involving good conservation practices, is essential for the livelihoods of many in developing countries. Existing policies often compromise this. Strengthening access regimes can offer opportunities for poverty eradication, as well as improving conservation and long-term sustainability. This national level action can be important for attaining globally agreed objectives, such as those in the MDGs, the CBD and UNCCD.

Secure entitlement refers to conditions under which users are able to plan and manage effectively. Secure access to natural resources can be an important stepping stone out of poverty, as it provides additional household wealth, which may support investments in health and education. Moreover, it may contribute to better natural resource management by supporting long-term vision that keep future generations and options in mind, and may encourage investment. Specifically addressing women's tenure rights is vital, as they play key roles in managing natural resources, and are particularly affected by environmental degradation. Intergovernmental initiatives, such as dam development, should not undercut local resource rights by shifting responsibility from the local to the national or regional level. To be effective, secure access rights may need to be complemented by addressing other barriers to sustainable and productive use, such as global trade regimes, insufficient access to capital and information, inadequate capacity and lack of technology. Valuation strategies, including payment for environmental services, can help ensure greater returns for local resource managers. Ensuring access to credit for small farmers and those directly reliant on ecosystem services is extremely important. Practice shows that finance schemes that especially target women can have higher than usual payoffs. Credit schemes, such as the Grameen banks in Bangladesh, can be designed to compensate those who ensure that environmental services are maintained. Improved local authority over natural resources can help diversify livelihood options, reducing pressures on resources that are under threat. Devolution of authority is one such mechanism. Despite a growing trend towards decentralization and devolution since the 1980s, and a broad policy commitment to give users greater authority, the institutional reform required to improve livelihoods is often lacking. Devolution needs to be complemented by capacity building and empowerment initiatives, improved tenure, and better trade and value-adding options.

Building and Bridging Knowledge to Enhance Coping Capacity

The roles of knowledge, information and education in reducing vulnerability converge around the learning process. The strengthening of learning processes for three specific objectives emerges as a key strategy to increasing coping capacity in a rapidly changing and complex environment.

Building knowledge about the environmental risk that threatens well-being, both among vulnerable communities themselves and among decision-makers at higher levels, is important. This involves both improved monitoring and assessment of the environmental, social and health-related aspects of pollution. It also involves mechanisms such as early warning systems (EWS) and indicators (for example, the Environmental Vulnerability Index) for communicating and disseminating information on environmental change. These systems should be integrated into mainstream development. One tool that has proved useful in this regard is poverty mapping. Poverty maps are spatial representations of poverty assessments. Poverty maps also allow easy comparison of indicators of poverty or well-being with data from other assessments, such as availability and condition of natural resources. This can assist decision-makers in the targeting and implementation of development projects, and the communication of information to a wide range of stakeholders.

Bridging knowledge for better decision making is also key. At the same time, the scientists and decision-makers should learn to listen to and to talk with these communities, and consider their unique, specialized knowledge that centres on human-environment relationships and the use of natural resources, even if it is not cast in the language of science. The most vulnerable should learn competencies and skills that enable them to adapt and cope with risks. The foundation for this and the learning processes above lies in a good basic education, as set out in MDG 2. This increases the ability both to understand information from public awareness and early warning campaigns about specific sources of vulnerability and to develop coping and adaptation strategies. For example, it was the poorest and least educated groups who did not heed the evacuation warnings for Hurricane Katrina in 2005. Educating the most vulnerable groups improves their coping capacities, and is also important for equity reasons. For example, the education of girls is one of the key means to break the intergenerational cycle of poverty. It is strongly associated with healthier children and families, and more sustainable environmental management.

Investing in technology for adaptation Science, technology, and traditional and indigenous knowledge are important resources for reducing vulnerability. Policies that facilitate the development, application and transfer of technology to vulnerable communities and areas can improve access to basic materials, enhance risk assessment practices and EWS, and foster communication and participation. Policies should support technologies that ensure equitable access and the safety of water, air and energy, and that provide transportation, housing and infrastructure. They should be socially acceptable in the local context. The opportunity lies in investing in a diversity of technologies, including small-scale technologies that allow decentralized solutions. Some types of technology can also be important resources for promoting social connectedness, stability and equality through democratization. Policies that facilitate communication, education and governance via information technologies, and that improve the status of underprivileged groups, are particularly valuable.

Developing countries stand to derive many benefits from technologies developed elsewhere, but they also face the greatest challenges in accessing these technologies and managing their risks. Commitments made in the Johannesburg Plan of Implementation (JPOI) remain largely unfulfilled. Computer and information and communications technologies, biotechnology, genetics and nanotechnology remain unavailable to vast numbers of people in the developing world. Past experience has shown the importance of attending to the appropriateness of technology's multiple connections with broader society, its fit (or lack thereof) in particular social, cultural and economic contexts, and its implications for gender. An important strategy to ensure this is to invest much more in capacity building in the countries for technology innovation and production. The UN Task Force on Science,

Technology and Innovation made a number of recommendations including: focusing on platform technologies, existing technologies with broad economic impacts (for example biotechnology, nanotechnology, and information and communication technology); providing adequate infrastructure services as a foundation for technology; investing in science and technology education; and promoting technology-based business activities.

Building a culture of responsibility The export and import of vulnerability is a recurring feature of the seven archetypes, meaning that many people – individually and collectively – contribute, often inadvertently, to the suffering of others while improving their own well-being. In this context, vulnerable communities need support to cope and adapt, so there is a need to build a stronger culture of "responsibility to act." Educating people about how their production and consumption patterns export vulnerability to other areas, continents and generations, and how this affects the prospects for living together at local scales, can contribute to a culture of responsibility. UNESCO's Education for All emphasizes the need to expand the view of education to include learning "life skills," such as learning "to live together" and learning "to be".

However, the chain of interactive drivers is far too complex to allow individual and collective actors to be aware of their own contributing roles and to feel more responsible to act. In addition, the institutional frameworks for addressing legal responsibilities to protect the global commons are often weak, particularly when issues cross international borders and happen over different time frames. A response strategy is needed where a culture of responsibility is based more on global solidarity for present and future generations as a way of integrating neighbourhood values with global solidarity. Such solidarity can be actively nurtured through, for example, education, processes of cooperative interaction, or the design of institutions that strengthen cosmopolitan aspirations and commitments.

Education for the purpose of learning to care for and feel empathy for neighbours, and through this build a culture of responsibility to act can be readily integrated into the overall strategy for both formal and informal education. Enabling learners to participate directly in environmental problem solving is one effective way to enhance conservation behaviour. Examples of teaching environmentally-relevant life skills include the education initiatives related to the Earth Charter and various programmes on global and world citizenship and human rights.

Poor governance, social exclusion and powerlessness limit the opportunities poor people have to participate in the decision making related to a country's resources and environment and how these have an impact on their well-being. Improved governance and tenure regimes may not work for the poorest people if the opportunities for their participation are not specifically strengthened. Improving opportunities for participation in governance and planning processes at local and higher levels of governance can help strengthen their coping capacity.

The UN Conference on Environment and Development (Rio conference) provided the basic institutional change for increasing participation in environment-related decision making. However, having a voice without being listened to and having an impact on outcome can lead to greater estrangement. Weaknesses in this aspect is a recurring complaint, for example, in the multistakeholder dialogues at the global level. Existing responses need to be strengthened, and active strategies to empower the most vulnerable could be developed by, for example, improving access to relevant environmental information, as provided for in Principle 10 of the Rio Declaration. This has already been implemented in many countries. Capacity building is also essential. Putting a strong focus on the equity aspects of the outcome of governance is another essential aspect of enhancing coping capacity and the legitimacy of governance. Equity-centred strategies involve identifying the most vulnerable groups and communities, assessing the impacts of suggested policies first and foremost on these groups, and taking measures to improve equity in access to resources, capital and knowledge.

Building capacity for implementation "Implementation failure" is common. There are many elaborate regional and global level multilateral agreements and action plans that have not been successfully implemented at the national level. The reasons behind the implementation failure are complex, and there are no simple solutions. Addressing this requires a multilevel approach. Three important opportunities can be identified: improving funding, investing in capacity, and developing effective monitoring and evaluation of existing plans and policies. International partnership is critical to success.

Increased financial commitment is essential to promote adaptation activities, increase human capability, support the implementation of MEAs and stimulate development. In developing countries, where financial resources are often constrained, creating better synergies between environment and development objectives is important. For example, there could be more interlinked health-environment strategies or poverty-environment initiatives. The incorporation of environment into Poverty Reduction Strategy Papers is one opportunity that can be more effectively used.

Official development assistance (ODA) continues to lag behind agreed targets. At the 1992 Rio conference, most countries pledged to increase ODA towards the UN target of 0.7 per cent of GNI. In 1993, the average level of ODA was 0.3 per cent of GNI. Describing the international redistributive system as is "in shameful condition," Brundtland emphasized that "the cost of poverty, in human suffering, in the wasteful use of human resources, and in environmental degradation, has been grossly neglected". The 2002 Monterrey Consensus recommitted developed countries to meeting the UN target. Since then, there has been a steady increase in aid, and by 2004 average ODA was 0.42 per cent of GNI. However, only five countries have met the UN target and by 2006 the average was down to 0.3 per cent again. The IMF's 15 richest member states have agreed to allocate at least 0.51 per cent of GNI by 2010, increasing this share to 0.7 per cent by 2015.

Investing in capacity building and necessary technology support, as envisaged in the JPOI and the BSP, can enhance ability to develop and implement required measures. Targeting capacity building at the right level is essential. Improved land management might require local capacity building, whereas addressing illegal movement of hazardous waste will require capacity building of the relevant agencies. In some areas, such as biodiversity management, capacity of some developed and developing countries to develop and implement interlinked strategies is lacking. Pooling resources, sharing best practices and collaborating in joint capacity building at the regional level have been successful.

Improving monitoring and evaluation capacity also hinges on increased investment in capacity building, and appropriate institutional and governance development. In some situations, there is a need for stronger government institutions, as well as national and international laws to ensure that standards are abided by. Better institutional and governance mechanisms, including measures to ensure access to relevant information and the courts, are necessary to support people in safeguarding their interests.

References

Barnett, J. (2003). Security and Climate Change. In *Global Environmental Change* 13(1):7-17.

De Soysa, I. (2005). Filthy Rich, Not Dirt Poor! How Nature Nurtures Civil Violence. In *Handbook of Global Environmental Politics,* P. Dauvergne (ed). Edward Elgar, Cheltenham.

Diehl, P.F. and Gleditsch, N.P. (eds.) (2001). *Environmental Conflict.* Westview Press, Boulder, CO.

Georges, N.M. (2006). Solid Waste as an Indicator of Sustainable Development in Tortola, British Virgin Islands. In *Sustainable Development* 14:126-138.

Homer-Dixon, T.F. (1999). *Environment, Scarcity, and Violence.* Princeton University Press, Princeton, NJ.

MA (2003). *Ecosystems and Human Well-being; a framework for assessment.* Island Press, Washington, DC.

6

Urbanization and Environmental Change

Urbanization is a global phenomenon that has transformed and continues to alter landscapes and the ways in which societies function and develop. Currently, more than half of the world's population lives in cities; the United Nations projects that by 2030 the world will advance to the 60% urbanization threshold. Rapid urbanization effects will not only be present within the immediate locations, but will be experienced regionally and globally. The UGEC project seeks to better understand these implications and the complex dynamic systems of urban areas that affect and are affected by global environmental change (e.g., climate change, natural disasters, loss of biodiversity, freshwater ecosystem decline, desertification, and land degradation).

The 7th Open Meeting of the International Human Dimensions Programme on Global Environmental Change held in Bonn, Germany showed the increased emphasis on the role of humans in global environmental change research – also recently acknowledged by the IPCC's AR4. A better understanding of the emerging challenges has sparked greater eforts towards research that includes a wide-range of disciplines, is policy relevant and pragmatic, and is communicated more effectively with governments, businesses, NGOs, and the general public. In this light, the conference addressed the current challenges associated with the human dimensions of global environmental change, focusing on the role of human beings, not only as contributors to, but also as actors able to frame problems and devise solutions for overcoming them.

Four social challenges – demographics; resources and technological innovation; social equity, cohesion, and sustainable adaptation; and adaptive institutions and governance – were identifed by the International Scientific Planning Committee as the most critical for sustaining healthy livelihoods and good entry points for discussing and advancing new research. With each topic as the daily theme running throughout plenary, parallel, and special session discussions, the human dimensions research community was provided a number of platforms to debate, share and take ideas forward, and to initiate new research collaborations. Sessions highlighted the interconnections that exist between specific issues of primary focus to IHDP such as water, land, food, climate change, coastal zones, institutions, technology or urbanization and one or more of these four social challenges. It quickly became clear that urbanization is increasingly becoming a high-priority issue for the human dimensions of global environmental change.

Cities as Agents of Global Environmental Change

Two independent (but related) processes working at the global scale are fundamental in our understanding of recent changes in governance of cities in the developing world. The first is globalization which implies economic fusion and the emergence of networks of world cities as the primary arena for action on global environmental change. The other is the drive towards decentralization, which aims to devolve more decision making authority at the local level, away from the national and regional levels. To a certain extent these processes are linked.

As Harris observes, "economic fusion, it seems, has stimulated political fssion" (in the form of decentralization). In his view, globalization is thus "liberation for cities.

While this view may be justifed for some cities in the developed world, for most cities in emerging economies the sovereign national state has been reluctant to relinquish its control of the local - particularly in the critical areas of city planning and fscal afairs. Tus, while the trend towards globalization has enhanced the image of high-growth metropolitan regions in developing countries as "global cities" - with the expectation of world class quality of basic services - this has not been matched with a concomitant increase in the capacity of urban local bodies to meet these expectations. The case of India - with its wide array of global cities - is particularly illuminating in this respect.

Capacity of Urban Local Bodies in India

A landmark event in the history of urban governance in India is the passage of the Constitution's 74th Amendment Act. The CAA left it up to the Legislature of a State to decide which powers and authority it may devolve on a municipality. In reality, most states in India have been reluctant to relinquish their control over fscal matters. Tus, although the CAA lays out a large range of local taxes, fees, and tolls to be levied by the Urban Local Bodies (ULBs), in practice the ULBs have imposed only a few of these. As a consequence, the ULBs have not been able to generate the resources needed to meet the requirements of their growing responsibilities.

Lacking adequate capacity to raise revenues on their own, city governments have largely been dependent on budgetary allocation from higher levels of governments. In recent years, however, large current account deficits at both central and state government levels are forcing city governments to look for alternative sources of funding through public-private partnerships and capital market funds.

A closer examination of these initiatives across diferent cities in India reveals that it is only the relatively large cities that have been successful in attracting this sort of funding. Ironically, while city governments are being pressured to raise their own resources for provision of basic services, they have very limited autonomy with respect to the planning and management of these services. Even for large cities like Mumbai, the planning and management of basic services falls under various ministries that have state-wide responsibilities and constituencies as opposed to an elected body of local representatives who are responsive to local needs. Phatak and Patel examine how this lack of capacity and autonomy at the local level impacted the recovery efort during the 2004 foods in Mumbai and how the response may have been diferent under a more decentralized form of governance.

Thus, to summarize, while the intention of the CAA was to move towards a decentralized system with municipalities functioning as institutions of self government, this has not been realized in practice. The degraded state of basic civic services and the rising frustration of the urban middle class have found instead a diferent outlet. This follows from the widespread global rhetoric on the potential of local citizen participation as a panacea to the problems of urban governance.

Citizen Participation and Urban Governance in Delhi

The metropolitan region of Delhi is referred to as the National Capital Region (NCR) of Delhi. It consists of the Union Territory of Delhi (also referred to as the National Capital Territory, or NCT of Delhi) and a few ring towns around it that fall under the jurisdiction of three diferent states shown. These ring towns were conceived as part of the Delhi Master Plan back in 1962 to reduce the population pressure on the city core. Although these ring towns have witnessed the highest population growth and have absorbed most of the immigrant fow, in terms of governance, these represent a major challenge. About 20-25% of the population in NCT of Delhi is estimated to live in slum and squatter settlements. Most of these people live in hazardous sites and are exposed to multiple environmental health risks.

Following the CAA and in response to the rising frustration of citizens with provision of basic services, the Government of NCT of Delhi launched a major initiative in 2000 called the "Bhagidari" (or partnership program) to empower the citizenry to look after local civic problems at the community level in collaboration with civic agency officials. The program created a formal system of liaison between residents, civil society organizations, and relevant government departments. The areas of consultation and sharing of responsibilities under the program include: rainwater harvesting, water and energy conservation, solid waste management, greening of Delhi, control of air pollution, and environment awareness campaigns.

As part of the program, several workshops are organized to train citizen groups and government ofcials to come together to discuss problems and brainstorm ideas in order to come to consensual solutions. Regular meetings are also held among the concerned departments and Deputy Commissioner Ofcers to implement and monitor solutions. Resident Welfare Associations (RWAs), formed at the neighborhood level, are the primary citizen groups under this program. Since the inception of the program in 2000, the number of RWAs in NCT of Delhi has grown from 20 to around 2000. In a city infamous for its bureaucracy and red tape, the RWAs now have a say in how the money is spent in their areas and how to hold the local ofcials accountable.

An important consequence of this program has been that it has moved at least the city's middle class from cynicism to active participation in local governance. Tus, the program plays an important role in bridging a perceived democratic defcit and has improved the accountability, transparency, and responsiveness of urban local bodies to the citizenry. However, it needs to be kept in mind that the Bhagidari program lacks statutory authority. An important criticism of the program is that it has created a parallel system of government, instead of strengthening the capacity of the existing structure of governance.

Another major problem is that the funding available under the Bhagidari program, even for small-scale community projects, is generally inadequate and the RWAs primarily depend on contributions from their members. This aggravates the underlying structure of inequities in the city, as the richer neighborhoods are able to collect enough funds to self-fnance community projects often at the cost of surrounding poorer neighborhoods. Decentralized governance under the Bhagidari system has meant that richer neighborhoods in these areas have been able to invest in deep borewells to cope with problems of inadequate and uncertain supply from the municipal system, while increasing the vulnerability of surrounding poorer neighborhoods and jeopardizing the sustainability of the system. Tus, to conclude, citizen participation in urban governance holds a lot of potential, particularly in terms of its role in filling up the democratic defcit. However, there is nothing intrinsically transformative about participation or about its pro-poor outcomes. The potential of participatory programs has to be seen in the wider context of institutional and political confgu-rations within which it is embedded, as elaborated above. Unless the capacity of elected urban local bodies is strengthened and their incentive structure is better aligned with the needs of citizen groups, participation by itself is unlikely to bring about any fundamental transformation.

Social Relations in the Waste Waterscapes

Political Ecology of Waste Waterscapes

In recent years, the urban environment has been addressed by a growing number of publications within the field of Urban Po-litical Ecology. Urban Political Ecology is grounded in an understanding of the city as a co-produced outcome of processes from the societal as well as the natural realm. Within this strand of inquiry, Swyngedouw has described socially contingent water realities in the city with the term "waterscape". Highlighting the waterscape's produced character allows for one to look into the historical trajectories and temporal dynamics which are shaped through negotiation processes between actors. Like the landscape, the waterscape may be "read" in terms of power struggles over the interaction that actors have with their environment. As such, waterscapes are understood to be highly politicised.

Tree factors, the distance between higher and lower ranks of the administration, the organisation at ward level, and the diversity of colonies translate into locally specific arrangements in which sanitation services are, or are not, being provided. As such, waste waterscapes are a localised phenomenon. Power in these waste waterscapes is decentralized, leaving the governance of waste water to be contested in multiple local struggles, a form of micropolitics. The outcome is a heterogeneous pattern of infrastructure, related practices and subsequent problems (Zimmer, forthcoming). Analysing these problems requires a focus on the people's interaction with the state through their everyday encounters with their representatives in the ward. Negotiating public service provision takes place within networks of actors including the administration, the political representatives and inhabitants.

Micropolitics in the Waste Waterscapes

As already noted above, ofcial sanitation approaches taken by Delhi's city administration do not necessarily translate to the reality on the ground. As sanitation services are organised at ward level, a certain amount of discretion allows the appointed DEMS staf to interpret rules and regulations. Moreover, the agglomeration of 16 million inhabitants is characterised by a heterogeneous pattern of colonies: while authorised areas house 23% of Delhi's population, slightly higher numbers live in squatter settlements - so called JJ clusters, (24%), and unauthorised colonies (UACs) (26%. Both are ofcially termed "slums", but while the demolition of JJ clusters is continuing with resettlement taking place at the outskirts of Delhi, UACs are in the process of being legalised.

Networks of Service Provision

In comparing a JJ cluster to a UAC, it becomes apparent that the networks of service provision are diferent for both types of slums. Service is provided to both slums by public scavengers. Inhabitants seeking to maintain and improve service delivery, therefore, interact with them directly. If this is not successful, slum dwellers have to use their links to other more powerful actors to "get the work done" by the scavengers. In both cases, addressing their political representatives is the most promising strategy for residents. Yet, while the UAC is in contact with the Municipal Councillor (MC), responsible for sanitation related issues in the ward under the MCD, the JJ cluster is linked to the Member of the Legislative Assembly (MLA), a representative at the State level. Although sanitation is not ofcially his duty, the MLA is actually related to the MC and so takes over some of his responsibilities. Moreover, inhabitants of the JJ cluster focus on addressing local staf of the city administration about service delivery, whereas members of the UAC are aware of a number of options to pursue that exist higher up the bureaucratic hierarchy. Like the confgurations of the networks, social relations within the networks are not the same. In the following, the interactions with scavengers and political representatives will be discussed.

Social Relations in the Network

While the network of service provision illustrates who is actively engaged in providing service to the slums, it is the social relations which determine how service provision is achieved. Negotiations of waste water governance take place through communication in everyday encounters. Tey can reproduce the status quo as well as challenge it. Against this background, the provision of waste water related services or lack thereof depends on multiple decentralised decisions within that network, with each interaction contributing to the fnal outcome. Like citizens, state employees can prove more or less powerful in these interactions.

In the investigated cases, the stability of power relations is not the same in both colonies. Inhabitants of the JJ cluster have a similar status to the scavengers in terms of education and income, which opens up the foor for a constant challenging of duties and rights. In the UAC, the status of inhabitants is signifcantly higher than that of the scavengers in terms of education and caste. Scavengers, therefore, face difculties in challenging the role assigned to them by the citizens.

A public scavenger working to pick up trash in the squatter settlement The opposite case presents itself concerning the relationship between political representatives and slum dwellers. While interaction is more fxed for inhabitants of the JJ cluster, it is more fexible for UAC residents. This is due to two factors. First, the JJ cluster represents only a small fraction of the MLA electorate, so that an inhabitant's ability to challenge the role of his or her representative is limited. In the case of the UAC, the whole ward is constituted of unauthorised areas; therefore, the MC has no other voters to rely on. Second, the social status, in terms of education and income, are strikingly diferent than those within the JJ cluster. This leaves slum dwellers with little power to challenge the MLA's infuential role. Meanwhile, in the UAC, the MC is a woman of the unauthorised colony itself, so that any diference in status depends exclusively on the powers her neighbours have vested in her. As her status depends on their goodwill, she cannot use her power as freely to deny inhabitant's requests.

Direct communication can be understood as a means to negotiate the described relationships, in terms of either reproducing or challenging them. Accordingly, forms of communication and the arguments used by residents to convince scavengers and political representatives are diferent in both settlements. In terms of direct communication with the scavengers, JJ cluster dwellers complain to them regularly and accuse them of not fulfilling their assigned duties. In this context, caste based reputations or prejudices are used by inhabitants to legitimise their claims to the scavengers. Moreover, inhabitants can put scavengers down referring to their job as "dirty". Although scavengers are members of the administration and are supposed to serve the citizens, they can argue in return that their government job gives them a better understanding of the waste water situation. Also, being employed by the state connotes a higher status than that of slum dwellers, referred to as "beggars" by scavengers. In the UAC, however, inhabitants rarely communicate with the scavengers directly, but rather complain to their political representative.

Concerning the relationship with politicians, it shows that with high power divides, inhabitants are reduced to supplicating (Corbridge et al., 2005) their representative to act in their favour or complain, if they have a higher social position. Residents of the JJ cluster, then, move between supplicating and complaining, using moral arguments more often than democratic ones. As constituents, this does put pressure on the MLA, but he is barely responsive to these arguments outside election times. In the UAC, residents more often complain to the MC, making democratic claims to their rights as citizens. She is regularly reminded through these arguments of the fact that she owes her post to them.

Politics of Space in the Megacity

The combined effects of rapid population growth, spatial expansion, environmental change and the complex local impacts of the global political economy, thus, go hand in hand with malfunctioning formal governance structures, spatial fragmentation, socioeconomic polarization and political struggles. Many of these conficts evolve around the “production of space”, in terms of the access to, use and appropriation of settlement areas, streets and market places.

In this context, the question of urban governance, its quality and spatiality, becomes an urgent one. The prevailing mode of governance is understood as a set of institutions, which actors in urban arenas accept as legitimate and concordantly align their actions, whether these are spatial practices or other social interactions. In our research, we combine such an actors-oriented governance approach with the notion of the “production of space” or “place making” in order to detect spatial connotations of urban governance. Tree key questions guide our analysis:

1. Who gets access to a certain space, for instance to a food market? (*access to space*)
2. Which spatial practices are applied by actors in these spaces? (*use of space*)
3. How and by whom is governance of space negotiated, contested and legitimated? (*appropriation of space*)

Our analysis mainly draws on case studies from ongoing empirical research into the megaurban food system of Dhaka. Here we present one which exemplifes the access to and the use of public space by street food vendors.

(In)formal Institutions, Conficts, and the “Production of Space”

In contested arenas many actors with divergent interests, unequal power positions and diferent capabilities constantly challenge the operating “rules of the game”. Given the ineffectiveness and often inappropriateness of state and municipal law, informal modes of governance, largely based on highly personalized networks, are often more efcient and pervasive. These sets of informal institutions manifest themselves in the structure of the local (not only “informal”) economy and in such spatial arrangements that are beyond recognition and – most often – beyond the control of state authorities. These arenas – understood here as the social “felds” in which actors are relationally positioned according to their diferent capacities and their respective positions of power – are constantly (re)produced by the actors themselves. Terefore, the rules that are in effect are often contested and highly dynamic; they can change not only over the course of years, but within a single day.

Specific modes of governance in Dhaka's food system, and in the megacity as such, go hand in hand with distinct spatial practices of actors. Henri Lefebvre claims: "(E)very society [...] produces a space, its own space." If we see informality as the hegemonic mode of urban governance or the major "organizing urban logic", then informal interactions create their "own space". Space is produced first and foremost through the contestations between hegemonic actors and subaltern actors (i.e. marginalized citizens, street vendors, etc.), the play of institutions, and in particular, through the perception of space. For instance, the encounters that hawkers have had with state authorities not only shapes the way in which they are "seeing the state" and formal institutions, but also how they are conceiving public space and thus, how they are exercising their "right to the city". So even though state or city authorities might wield power over "their" citizens and formally regulate the city territory, they can hardly control (all) the social, cultural, political and economic (inter)actions and imaginations of the city dwellers. This inherent agency of actors, though often limited, can lead to everyday spatial practices in resistance to the formal institutions set by the state, an effect that Bayat has coined the "quiet encroachment of the ordinary". The following case study reveals daily negotiating processes of obtaining access to and using urban space from the perspective of street food vendors.

Street Food Vendors and the "Quiet Encroachment of the Ordinary"

In Dhaka, just like in Hyderabad, Bogota, Mexico City or Lagos, the sale of food along streets and in public places is, at present, a highly contentious issue. While street food vending is a good self-employment opportunity for the urban poor and an important functional element in urban food systems in countries of the Global South, it is seen as obsolete, unhygienic, disorderly and "in the way" by city authorities and planners. Nevertheless, everyday vendors claim the streets, consciously ignoring formal rules.

Street food vendors apply diferent spatial practices according to the types of food they sell, their social and economic capital, their perception of space, and their perception and understanding of the operating rules at their particular vending site. Consequently, there is a great range of vending styles that can be subsumed in four broad categories. Permanent shops have been solidly built, usually illegally, often located in small niches right next to a footpath. Semi-permanent vending units are larger tables that are set up for the day at the side of a footpath, but can be dismantled quickly and are usually removed at night. Semi-mobile vending units, such as push-carts or rickshaws, can be moved to reach consumers at diferent places at diferent times; however, they more often remain at one site for the whole day. Lastly, mobile vendors sell their products by walking around with a tray or basket of fruits, snacks or a fask of tea; they squat on footpaths and occupy parts of roads spread out over public places.

These vending styles can be considered diferent expressions of the production of space. As "the issue of space is simultaneously an issue of order and ultimately an exertion of power", state actors are compelled to react. The mobile vendors do not threaten the state's monopoly in regulating public space; therefore, they are largely tolerated. In contrast, the more permanent vendors not only challenge the state's authority through their appropriation of space, but they also question the state's formal institutions by making their own operating rules for their respective vending sites. Semi-mobile, semi-permanent and permanent vendors are, thus, the major targets of evictions. These (often violent) "clean-up-drives" are at present, more frequently carried out by the police in order to improve "public security" and "food safety" in Bangladesh, and thereby demonstrate state authority. In turn, the vendors have to adapt to the erratic state's eforts to re-claim the streets, so as to reduce business loss and secure their livelihoods in the longer run. While the mobile and most semi-mobile food hawkers simply move their vending units quickly when they perceive a threat, the more permanent vendors hide their food and vending equipment near their vending site. However, many vendors wait-and-see and then draw on their social capital to reduce their loss, bridge income shortages and reclaim "their" vending site, for instance by bribing the police or seeking help from local power brokers (mastaans). Knowing "the right people" is, therefore, not only the key to getting access to a specific vending site, but this also determines the vendor's use of space and the extent to which the vendors can participate and shape the "politics of the street".

Urban Governance and Modern Challenges

Governance systems must be able to accommodate change and adapt quickly to changing environments. However, cities everywhere face the challenge of adapting water management strategies within governance systems that lack the fexibility needed for innovative change.

One entry point from which to understand and evaluate adaptation in a comparative urban governance context is to examine the role institutional features play in shaping policy outcomes. At the city level, institutional features such as the type of government, utility provider, and election rules have been shown to infuence decision making in urban policy areas such as solid waste recycling, urban redevelopment programs, and water pricing. Strong executives can drive policy agendas and avoid gridlock and debate, providing momentum to selected causes and not to others. The type of utility – whether a specialized entity, privatized provider, or embedded within general government – can signifcantly infuence the incentives, resources, and planning horizon of decision makers. Election rules for city ofcials such as councils, mayors, and board members can radically shift constituencies of elected ofcials and their accessibility for specialized interest groups. The degree to which these institutional features are fragmented among a metropolitan area has been shown to shape not only policy outcomes, but the form of cities themselves. Inconsistent and conficting authority inhibits coordination and fosters localized, myopic decision making.

However, the infuence of institutions on urban policy outcomes goes well beyond the formal structures of urban government. New developments in urban political economy can also drive policy outcomes. For example, the growing infuence of the private sector in both urban and water policy can introduce new options for urban water services. Technological advancements such as water recycling, desalination, and reverse osmosis and nano-scale water membranes are often driven by private sector investment and research. Australia has several mechanisms for water businesses to access foreign markets and clients, with a strategy of marketing the "Water Brand Australia". With the emergence of "global" cities comes diferentiated urban policy, increasing strategic action at the national level, and more direct network links between cities and the international arena. Combined with the growing authority of environmental concerns in many places, these features of urban governance create new and diferentiated institutional effects on policy outcomes.

Water Recycling

Water recycling – also referred to as water reclamation and water reuse – is "the treating and managing of municipal, industrial, or agricultural wastewater to produce water that can be productively reused". The existing and planned use of recycled water in urban areas is quickly accelerating. Governments, utilities, communities, businesses, and non-governmental organizations (NGOs) around the world have begun to weigh in on the advantages, disadvantages, and future role of this relatively new and sometimes controversial innovation in urban water management. Furthermore, water recycling requires deliberate investment in new technology and infrastructure through the coordination of multiple aspects of urban water management (e.g., wastewater, water supply, land use, infrastructure investment, and public perception), making it both a salient and analytically useful case to examine the infuence of institutions.

California and Australia have been leaders in the area of water recycling, but have had diferent challenges and successes. Both places are well-developed, urbanized, democratic political units in the midst of drought; both also have highly populated arid areas and less populated wet areas; in both places decision making authority for water management lies at lower levels; and both places have fairly dichotomous urban and agricultural development. In California, authority is located with local water suppliers and in Australia, authority over water resources is constitutionally vested in states and often exercised by state-owned water companies. Tey are two of four governments worldwide that have regulations for water recycling, and they have developed recycled water for a variety of purpose. In both cases, their central governments have made water recycling a priority for the future sustainability of their respective regions.

Despite these similarities, Australia has been more successful than California in developing recycled water for household use in urban areas. Cities in Australia have begun to incorporate recycled water into residential water supplies: water suppliers provided 1.8 gigaliters of recycled water to households in 2005, and 16% of Australian households recycle water. California has not done so, despite greater investment and capacity for water recycling and a strong focus on urban water supplies presented in its most recent water plan. A signifcant advantage for Australia in terms of incorporating recycled water into residential supplies may be its institutional characteristics.

Variations in institutional features may be responsible then for these diferent successes. Water supply and sanitation in Australia are managed jointly through coordinated agencies, and the country has forged a number of intergovernmental agreements to coordinate the actions of the states with the goals of the federal government. The water business sector in Australia is also highly organized and is generating more afordable recycling options and developing better networks with urban water managers. In California there are separate agencies responsible for each part of the urban water system and coordination between the state government and local water suppliers are weak. Australia's highly coordinated system may reduce transaction costs in decision making, but may also limit the number of access points for public input.

Intergovernmental coordinating mechanisms are also stronger in Australia, further reducing institutional complexity and associated transaction costs. In Australia, each state has signed on to the National Water Initiative (NWI) of 2004. This is Australia's "blueprint for water reform", and outlines priorities, targets, and actions for water management authorities throughout the country. Tus, while the Commonwealth government does not have constitutional authority to dictate what these priorities, targets and actions will be, they have been able to successfully establish these jointly with state governments. The effects are felt throughout state agencies, academia, and local organizations in part through the large sums of money mobilized by the NWI, but also because states have internalized the standards into their own policies and practices. In California coordinating mechanisms have not been as successful in terms of establishing a shared vision for water reform. Two prominent examples of this are the State Water Plan and the Best Management Practices developed by the California Urban Water Conservation Council. While establishing priorities, targets and actions that water agencies should take to improve economic and environmental outcomes, the coordinating effects have not been signifcant. There are fewer economic incentives and less buy-in from water and wastewater agencies.

Countries around the world are looking to expand their water recycling capabilities. The complexity of water institutions and the incentives they generate will help determine the path these projects take and whether cities will ultimately face less risk in the future. Coordinating levels of decision making through water supply and wastewater management will be an important first step in many places. This could take place through consultative bodies, joint agreements, or full institutional reform. An emphasis on local resources meeting local needs will help to incentivize water reuse and capitalize on place-specific expertise, and allow for a democratic decision making process. However, water management in many places is highly decentralized, which can present challenges due to the wide range of local level capacity (Dinar et al., 2005). Terefore, it is critical that institutions are able to both counter fragmentation and capitalize on site-specific expertise by reducing transaction costs, fostering shared ideas and norms, and encouraging experimentation.

As cities face the challenges of the coming decades, sustainable urban water management will become a necessary part of planning and policy. Urban governance institutions will help shape our response to these challenges, but should be understood in the context of broader developments and relationships. Water recycling is one example among many of the kinds of water management options cities are grappling with to adapt to changing demographic and climatic conditions. Identifying key institutional features that can be adapted or reformed is an important step toward meeting urban water challenges. Further research into the role of institutions will help illuminate not only why we observe the outcomes we do in diferent cities around the world, but also the mechanisms through which these outcomes may be infuenced to enhance sustainability.

Urban Transitions for Progressive Climate Proofing

Urban settlements are being recast as hotspots of risk with calls for urban development to be climate proofed in response to both adaptation and mitigation. This, in turn, requires that the capacities for transition to a climate proofed future are made clear. To date, the integration of climate proofing into urban policy has tended to focus on legislation and technical innovation for mitigation such as transportation planning, architectural design, clean air and solid waste management that can help mitigate hazard. There has been much less attention on adaptation planning and the cultural, social, and political contexts that determine policy and action. Resilience theory offers two useful mid-level theoretical tools for evaluating capacity for transition: social learning and self-organisation. Social learning (the possibility that individual innovations can spread throughout social systems) and self-organisation (the possibility for spontaneous and unplanned collective action) are believed to be key attributes for social systems that can identify and respond to unexpected shocks such as those associated with global environmental change. Beyond this, the appropriateness of resilience is constrained by its emphasis on systems persistence. The core functions of dominant social systems, social control to enable the conversion of labour and environmental assets into economic wealth, rarely serve the interests of the majority population or equitably meet those of minority groups so that their persistence is not automatically a desirable attribute. Core functions are perhaps most contested in urban systems where freedom from direct dependence on ecological resources and the presence of social and cultural diversity mean there may be few shared core needs and values.

Taking social learning and self-organisation forward, work on transitions ofers a meta-theoretical framing and one that points analysis at the interaction of ideas or collective action over scale as well as space and time. Transitions are processes of profound change affecting institutions, structures and actor behavior and values, so that a post-transition system operates according to new assumptions, rules and practices. Geels argues that it is at the local level that symptoms of crisis are identifed and may be responded to promoting transition; the regional level is resistant to change due to internal alliances and linkages; but, the national and global level can be a source of incremental transition through changing economic policy or environmental conditions.

Social Capacity for Progressive Climate Proofing

The Mexican Federal State has a good track record on climate proofing (as far as any national position can be so described) including a National Strategy on Climate Change launched in 2007 and preparations for a Special Programme on Climate Change. Tanks to these eforts Mexico has jumped from 14th in 2006 to fourth in 2008 out of 56 countries ranked according to their climate change performance in the Germanwatch "Climate Change Performance Index". By contrast at the regional level, Quintana Roo has not integrated climate proofing into development policy, although it does take the associated domain of hurricane risk management seriously, and has a good record in containing human loss. The most recent event, Hurricane Dean, in 2007, caused limited human impact.

Within this wider context, the data presented below allows a comparison of capacity for transition in two towns from the viewpoint of civil society. Mahahual is a pioneer settlement with a population of about 500 largely in-migrants from Mexico and internationally. The local economy boomed with the construction of a cruise ship terminal but has been in decline since this was closed as a result of Hurricane Dean. Playa del Carmen has a successful and growing economy based on international tourism and in 1994 became the capital of the newly created Municipality of Solidaridad. In 2005 its population exceeded the 100,000 inhabitants. Playa has experienced direct hits from hurricanes. The worst challenge came in 2005 with category 4 hurricanes Emily and Wilma; just a few months apart. However, there were no fatalities and the town recovered very rapidly. In fact, the local tourist economy benefited from the relocation of tourists from Cancun which had been hit even harder by Wilma.

The joint resilience-transition framework highlights the limits to bottom-up climate proofing within the contemporary development regime experienced in each study site. Importantly, state level institutions were critical in blocking incremental transition from the federal level and also in constraining local capacity for action. In both sites this was compounded by values oriented towards individualism, short-term economic proft motives and a lack of identity with place, which served to constrain discourses on climate change and development and visions of what might be possible for local social action. This combination of resistance in formal institutions at the state level and the abdication of individual responsibility for action justifed through dominant local discourses of development is surely a considerable constraint on capacity for progressive climate proofing.

Social learning was inhibited in both Mahahual and Playa by a lack of access to scientific knowledge on climate change and climate proofing options. In our interactions with local agents, climate change was discussed but mostly indirectly, through local development challenges – beach erosion, loss of urban green space and solid waste management in Playa, and loss of mangroves following Hurricane Dean and post-Dean reconstruction in Mahahual. Responsibility for behavioural change to mainstream adaptation and mitigation was largely pushed onto the state or to an anonymous global community, reffecting strong centralisation in Mexico and Quintana Roo and the alienation of local actors from responsibility for climate change action. Dependency on paternalistic state structures was questioned in the face of state inaction on climate change. Yet, successful evacuation following Hurricane Dean was presented as proof of state competence. In this way demonstrable state capacity in disaster response was extended to equate with climate change, and allowed respondents to abdicate responsibility for climate proofing. Certainly, climate proofing had not become a rallying point for alternative visions of development. Scope for institutionalisation of mitigation outside of the state was identifed in Playa through family run hotels with stakes in sustainable development, but overall respondents viewed the role of private sector activity as focused on fast economic gain with little concern for social development or climate proofIng.

Social organisation in Mahahual was greatly undermined by a lack of trust in organised social action and in community leadership. Against this background, Hurricane Dean was cited as having catalysed collective action through installing a shared sense of place amongst residents who stayed behind after the hurricane to rebuild. Playa's small, professional but competitive civil society undermined scope for coalition forming. Underlying this was a weak cultural base – individualism was the norm with little formal culture of volunteerism and a sense of place was limited. Even though many migrant residents had put down roots, they did not self-associate with Playa. Government was constrained by a highly competitive and dynamic political culture with politicians and ofcials coming in and out of ofce frequently and seldom building on past knowledge or initiatives. In both settlements, outside skills and information were successfully drawn on by civil society actors.

Given the constraints observed for both sites, any bottom-up civil society led capacity building process for progressive climate proofing in Quintana Roo is likely to be challenging. While innovation may arise from local actors, the scope for up-scaling or horizontal replication is limited by resistance at the state level and in common culture. Consequently, in the current context, change at the scale and pace required to meet the challenges of adapting and mitigating to climate change requires top-down, government led action. Action will need to engage with local actors for legitimacy and appropriateness, and collaborate with the corporate private sector, but nonetheless needs federal and state level government leadership. The need for concerted action across governance scales to reform the institutions of state government and challenge dominant culture is perhaps the most important outcome of the resilience-transition framework. This need is likely to be shared by those many societies that are exposed to climate change risks, but have not yet integrated climate proofing into development planning.

Managing Urban Growth in the Context of Environmental Changes

Although industrialization, the service sector, infrastructure, communication and trade have grown, rapid urbanization results in a disproportionate growth in population against the corresponding economic transformation. The consequences of such rapid urbanization are largely negative including the alarming increasing incidence of poverty, urban sprawl, inadequate social services and infrastructure defciencies. The negative consequences have subsequently led to the deterioration of conditions in human settlements, pollution, and the depletion of natural resources.

Due to the pattern of urban growth and other factors, African countries have chronically faced the challenge of managing urban development and still grapple with mobilizing fnancial resources, reforming urban governance, mobilizing communities, investing in urban infrastructure, sustaining the urban environment and providing social services; all of which would promote the necessary growth of urban economies sufcient for tackling urban poverty. Africa is projected to have the world's shortest urbanizing period and as it enters the urban "age" it will also be affected by global environmental change and most markedly, climate change. Africa is the most vulnerable to climate change because of its limited capacity for recovery and management of disasters. Africa is already experiencing extreme events in terms of climate variability and climate change impacts from droughts, fooding, violent storms and sea level rise, all of which have put many cities at risk and made millions of Africans vulnerable. Tus, Africa faces two main challenges at present; first, is its speed of urbanization - high compared to other regions, but Africa lacks the needed corresponding speed of response for such a challenge. Second, the vulnerability of urban areas to climate change impacts.

Drivers of Urbanization and Development Challenges

Several drivers are responsible for urbanization including population dynamics, economic growth, legislative designation of new urban centers and increases in densities of rural trading centers. Notably, the megacities such as Lagos, Cairo and Kinshasha are expanding further. Although urbanization presents opportunities in sub-Saharan African countries, the challenges for urban development are overwhelming. These challenges include social service provision, sustainable economic development, housing delivery, urban governance, spatial development guidance and environmental management for disaster and risk reduction. In addition, these challenges pose sustainability concerns in social, economic, environmental and institutional contexts, requiring well designed strategies to create positive impacts of urban development.Sustainable urban development and management should be of high priority in sub-Saharan Africa. As observed by UN-Habitat, systemic policy failure has provided ground for the proliferation of informal cities. Since urban development is crucial to social transformation as engines of growth, urban centers need to be managed properly in order to enhance and promote regional development. With the emergence of urban corridors across the region, there is hope that urban areas will play a signifcant role in unlocking rural lands and guiding population away from the primate cities.

Climate Change and Implications for Sustainable Development

As the process of urbanization accelerates in Africa, there is growth of an "emerging" economic sector that not only contributes to national economies but also provides employment and livelihood for many urban dwellers. However, this emerging sector together with other urban sectors, are facing the wrath of environmental change and particularly climate change. Climate change concerns are intensifed by the alarming growth of the slum population across the region. As observed by UN-Habitat, slum population growth is averaging between 3-6% and in many countries it is higher or equal to urban population growth. This raises concerns for urban sustainable development in the context of environmental change.

Attention is increasingly focusing on the societal conficts associated with environmental sustainability; conficts that become more evident as urbanization levels increase while urban environment degradation accelerates. In the developing world, environmental considerations are continuously challenged because they have not been coupled with an efcient programme of easing urban poverty. Improvements in the urban infrastructure, education and the health care service, housing, transportation and communication cannot be made without giving equal rights to the poor and the other disadvantaged social groups, and without recognizing their informal organizational capacities. No environmental policy can be efcient without a social policy. Capacities of communities to cope with environmental challenges with or without support from the city authorities have been described as informal forms of organization, for example, the self provisioning of infrastructure and services. Sustainable urban development can be looked at as encompassing six areas: (1) governance, (2) social and cultural considerations, (3) social infrastructure and public services, (4) urban land use and housing issues, (5) urban transport and urban natural resource management, and (6) employment and the enhancement of economic growth. These six areas form the framework for understanding the sustainability of urban development and also shed light on how to achieve sustainable urban development.

The Case of Uganda

City management in sub-Saharan Africa is characterized by a mix of policies and approaches including decentralization, centralization and autonomous authorities. Although an effective urbanization policy is one of the several instruments to guide development in many of these countries, there are disproportionate policy responses resulting in inconsistencies and unguaranteed orderly and sustainable urban development. For example in Uganda, the current Local Governments Act has succeeded in the devolution of power to local governments for effective planning and delivery of services to the urban populations. Although the Act was implemented in 1993 and is still possibly in its gestation period, analysis of its efciency thus far has indicated shortcomings in the requirements for sustainable urban development. One lesson to take from Uganda is that challenges of urban development include the effective and efcient provision of urban services, governance, infrastructure development, climate change mitigation and adaptation. These are key aspects of urban development not only because of their infuences on the economic, social development and environmental status of an urban area but they are also necessary due to increasing climate change vulnerability. Social and economic development needs a spatial framework but decentralization has led to competitive allocation of resources with little consideration of equitable spatial development of the city. One of the key consequences of this is marginalization of large sections of the urban population, leading to polarization.

Furthermore, there are many risks associated with environmental change in sub-Saharan Africa including: urban health, urban transportation, urban energy consumption, urban water and sanitation and urban waste management. For example, in Uganda energy demand stands at 0.00012 ton/vec.km for the public transport sector. And, while improved drinking water access only rose to 54% in 2006, with sanitation improving to 29% in the same year, garbage collection coverage in one of the municipal divisions increased from 10% to 80%, reducing the unit cost (in Uganda shillings) of collection from 11,300/m to 4,500/m; or the equivalent of US$8/m to US$3/m. With the level of energy use coupled with poor management of wastes, the burden is placed on the environment through GHG emissions. It is therefore necessary for urban sustainability to be placed.high on the development agenda in sub-Saharan Africa. Other challenges that face these nations include the increasing social polarization of urban communities, environmental degradation, increasing burdens on large proportions of the urban populations (i.e., poor sanitation, fooding, wastes accumulation, deteriorating public health, and disasters), and regional imbalances in urban development. Urban areas striving for sustainability will have to address the increase in urban populations, industrialization and the associated demand for housing that ushers in problematic processes of land use/cover changes; the nature of urban expansion and extension around major urban areas and peri-urban areas that creates severe social and environmental consequences; and the increases in exposure and sensitivity to natural disasters in urban areas.

In conclusion, despite potentials for realizing sustainable urban development, such development generally remains elusive in sub-Saharan Africa. Social sustainability of urban areas would require adequate cross-sectional distribution of urban services to ensure accessibility and improve the conditions of many urban dwellers. The challenges posed by urban growth and expansion to sustainability are daunting by far, given the current experiences of populations with inadequate access to urban services including water supply, sanitation, education and health; while inadequate infrastructure such as drainage systems and roads are exposing urban populations to environmental burdens. The consequences of these conditions are increasing urban poverty and urbanization of poverty. The intertwining of urban poverty and environmental challenges calls for innovative urban management supported by research. This will require a concerted efort that should bring together researchers, policy actors, communities and governments to address such challenges.

Urban Vulnerability and Adaptation to Climate Change

Nigeria is highly vulnerable to the impact of climate change on many fronts considering its geography, climate, vegetation, soils, economic structure, population and settlement, energy demands and agricultural activities. Its location, size, and geographic relief give rise to a variety of climates ranging from the tropical maritime climate of the rainforest located along the coastal and southern sections of the country, to the tropical hinterland climate of the Sahel in the northeast. Nigeria has a population of about 140 million that impacts the physical environment through various activities within an area of 923,000 square kilometres. Sixty percent of people live directly on the natural resource base as farmers, cattle rearers and fsherfolk, while the informal sector constitutes the bulk of the urban population. In all, the level of adopted technology is rudimentary, leading to low output and high levels of poverty. The institutional capacity to manage the urban crisis is also very low. Consequently, the impact of the human socioeconomic activities on the environment is devastating. Climate change has led to a shift in the boundaries of major ecological zones which has resulted in the heightening of drought and desertifcation in the marginal arid zones of the country; aggravation of soil erosion and fooding in areas of higher rainfall; and salt water intrusion along the coastal belt. Climate change has also altered animal and plant composition in diferent regions of the country.

Successful national economies depend on well-functioning and resilient urban centres. It is therefore crucial to safeguard the cities of Nigeria against vulnerability to climatic changes, so as to become catalytic to national development goals. For the most prosperous and well-governed cities, adaptation to the likely risks from climate change over the next few decades does not appear problematic. Such action requires adapting buildings and infrastructure to the increasing risks; working with population groups and settlements most at-risk to fnd solutions that best serve them; and good disaster preparedness. But in Nigeria, as in other developing countries, one cannot adapt infrastructure that is not there. The vulnerability of low-income urban dwellers to climate change is often ascribed to their poverty – but it is far more the result of failures and/or limitations in local governance to ensure that needed infrastructure is in place. In addition, the economic opportunities available for people to choose are very narrow and are mostly limited to harvesting the products of nature with rudimentary technology. The institutional capacity for recognizing the enormity of the problem of climate change, enacting appropriate legislation as well as enforcing it is also severely limited. Furthermore, the availability of climate change information is very low and attitudes towards environmental degradation are also very poor. It thus becomes important to integrate adaptation strategies into both municipal and national development plans.

Against this backdrop, the major questions that must be addressed in Nigeria are:

- What are the factors contributing to climate change vulnerability and what are the possible adaptation strategies?
- How can municipal governments in Nigeria be strengthened to effectively plan adaptation to climate change?
- What are the investment options for adaptation programmes and the costs associated with them?

Tese, among others, are the issues and challenges that must be urgently addressed for climate change adaptation to take place in Nigerian urban areas.

Chronic Vulnerability in Niger Republic

The Sahel Region is one of the poorest and most environmentally damaged places on earth. Low productivity per land unit is a challenging feature of most dryland ecosystems. Limited rainfall inputs, small and highly localized surface water resources, scant or non-existent vegetation, and low population totals all combine to place constraints on the total productive capacity of both the cultural and physical components of dryland ecosystems. The biggest security threats faced in Africa and in particular, the countries in the Sahel Region of Africa, are economic and social threats including poverty, infectious diseases, and environmental degradation. Most importantly, ecological disaster has become a serious concern to the Sahelian people, particularly in light of climate change impacts. These are concerns of immense importance when one is to consider the preparedness of the Sahel Region for fast-paced urbanization.

Niger Republic, a Sahel country, ranks 177 out of 177 countries in the United Nations Development Programme Human Development Index, thus, ranking it as the poorest nation in the world. The country sufers from endemic poverty, as over 60% of the country's 11.5 million people live on less than one dollar per day. In spite of some progress within the health and education sectors, progress towards the UN Millennium Development Goals is still slow and difcult. In 2004, a severe drought and subsequent locust invasion destroyed up to 100% of the crops in some regions, causing food shortages for more than one third of the population. According to a World Health Organization (WHO) report, an estimated 800,000 children under five years old sufered from hunger, of which 160,000 were moderately malnourished and 32,000 severely malnourished. The most affected were the poorest and most vulnerable agro pastoral regions of Tillaberi, Tahoua, Marah, Difa, Agadez, Zinder and Gaya. In these regions, pregnant and lactating women, totaling 261,300, and children were most at risk. Furthermore, the majority of Nigerien families are subsistence farmers and livestock breeding plays a key role for these agro pastoralist families. Any loss of livestock or decreases in the market value deprives the household of its major resource and exposes it to food insecurity. Many actions have been taken towards reversing the effects of degraded lands, but the unknown mechanisms of desertifcation encroachment can impede results.

Today, there are many strong forces transforming the world; environmental and economic forces are rearranging relations among nations. As the process of globalization accelerates, the United Nations will fnd itself confronting many of the world's challenges. United Nations agencies specializing in humanitarian eforts have provided relief materials to the victims of ecological disaster in Niger. However, the emergency programs were limited to the provision of relief food, resettlement of destitute pastoralists and sometimes, salvaging livestock. Huge sums were spent on imported relief food to feed drought-aficted pastoral peoples and other rural communities in Niger Republic, but more attention was given to food supply rather than health risks associated with malnutrition – such as the increase in death from disease outbreaks among people that had been weakened by malnutrition. It is imperative to note that the excessive use of relief food creates dependency and erodes people's initiative to support themselves.

References

Anderson, K. (2004). Subsidies and Trade Barriers. In *Global Crises, Global Solutions* (ed. B. Lomborg). Cambridge University Press, Cambridge

Aron J.L., Ellis J.H. and Hobbs B.F. (2001). Integrated Assessment. In *Ecosystem Change and Public Health. A Global Perspective,* Aron, J.L. and Patz, J.A. (eds.). The Johns Hopkins University Press, Baltimore and London

Bhagwati, J. (2004). *In Defense of Globalization.* Oxford University Press, Oxford

Chambers, R. (1995). Poverty and livelihoods: Whose Reality counts. In *Environment and Urbanization* 1(7):173-204

Solecki, W.D. and Leichenko, R.M. (2006). Urbanisation and the Metropolitan Environment: Lessons from New York and Shanghai. In *Environment* 48(4):6 – 23

7

Forest Processes and Environmental Change

Studies of the effects of climate change on forests have focused on the ability of species to tolerate temperature and moisture changes and to disperse, but they have ignored the effects of disturbances caused by climate change. Yet modeling studies indicate the importance of climate effects on disturbance regimes. Local, regional, and global changes in temperature and precipitation can influence the occurrence, timing, frequency, duration, extent, and intensity of disturbances. Because trees can survive from decades to centuries and take years to become established, climate-change impacts are expressed in forests, in part, through alterations in disturbance regimes.

Disturbances, both human-induced and natural, shape forest systems by influencing their composition, structure, and functional processes. Indeed, the forests of the United States are molded by their land-use and disturbance history. Within the United States, natural disturbances having the greatest effects on forests include fire, drought, introduced species, insect and pathogen outbreaks, hurricanes, windstorms, ice storms, and landslides. Each disturbance affects forests differently. Some cause large-scale tree mortality, whereas others affect community structure and organization without causing massive mortality. Forest disturbances influence how much carbon is stored in trees or dead wood. All these natural disturbances interact with human-induced effects on the environment, such as air pollution and land-use change resulting from resource extraction, agriculture, urban and suburban expansion, and recreation.

Each disturbance has both social and economic effects. Estimating the costs of each of these disturbances is very difficult; these estimates for the United States are illustrative only. Of the eight forest disturbances considered, ice storms are the least costly, averaging about $10 million and more than 180,000 ha annually. Insects and pathogens are the most expensive, with costs exceeding $2 billion and 20.4 million ha per year. The socioeconomic aspects of these damages are only part of the cost. Costs of impacts to ecological services (e.g., water purification) can be large and long term.

Past and Future Climates

The Earth has experienced cycles of temperature and precipitation change on a geological scale, but recent evidence points to a large anthropogenic component to current global climate changes. Analyses of the last 100 years of climate data for the coterminous United States suggest that the average temperature has risen by 0.5°C and that precipitation has increased 5%-10% ; observations also indicate that there has been some increase in precipitation and temperature extremes. To look at future climates, scenarios from two of the newer, transient general circulation models (GCMs)—one developed by the Hadley Center in the United Kingdom (HADCM2SUL) and one by the Canadian Climate Center (CGCM1)—have been selected for this national assessment. These transient GCMs simulate atmospheric dynamics under a gradual increase in greenhouse gas concentrations from about 1895 to 2100 and produce scenarios (precipitation patterns, temperature changes, and so on) that forest-process and biogeography models use to examine transient community and ecosystem dynamics under climate change.

These two climate scenarios present a useful contrast for future climates. The HADCM2SUL produces relatively modest temperature increases over the United States (approximately 2.6°C) and large precipitation increases (about 20%); the CGCM1 simulates larger temperature increases (approximately 5.0°C) and similar model precipitation increases over the coterminous United States in the next 100 years. The ecological models associated with the national assessment incorporate fire and drought disturbances, and we report the implications of these climate scenarios on these disturbances. The technology to incorporate other disturbances, such as windstorms or invasive species, is only now emerging. Therefore, the analyses we present here are based on new technology or are simply our best inference based on ecological models, literature surveys, or our professional judgment.

Climate Influences on Forest Disturbances

A review of how each disturbance is influenced by climate, affects forests, and might be exacerbated by climate change provides a background for examining ways to cope with the impacts of climate change. The effects of each disturbance are partly tempered by prior adaptations. For example, species present in a forest reflect past disturbances. Droughty sites typically support species that survive well under dry conditions with uncertain rainfall. Sites that have frequent fires contain gymnosperm species with serotinous cones. Thus, if climate change alters the distribution, extent, frequency, or intensity of any of these disturbances, large impacts (such as loss of species regeneration) could be expected. The effects on species or communities already at the margin of their range may be particularly severe.

Fire

The frequency, size, intensity, seasonality, and type of fires depend on weather and climate in addition to forest structure and composition. Fire initiation and spread depend on the amount and frequency of precipitation, the presence of ignition agents, and conditions (e.g., lightning, fuel availability and distribution, topography, temperature, relative humidity, and wind velocity).

Fire effects on forests include acceleration of nutrient cycling, mortality of individual trees, shifts in successional direction, induced seed germination, loss of soil seed bank, increased landscape heterogeneity, changes in surface-soil organic layers and underground plant root and reproductive tissues, and volitalization of soil nutrients. Erosion can occur where soil disturbance accompanies fire (e.g., during fire fighting or timber salvage operations). Fire affects forest value for wildlife habitat, timber, recreation, and, through smoke, human health.

The rapid response of fire regimes to changes in climate can potentially overshadow the direct effects of climate change on species distribution and migration. Modeling results predict great variation in future fire-weather patterns for the northern portion of North America. The seasonal severity rating (SSR) of fire hazard increases over much of North America under both the HADCM2SUL and the CGCM1 scenarios. The wetter Hadley scenario produces some small decreases in SSR for the Northern Great Plains, and increases are generally less than 10% over most of the rest of the continent. Some fire history studies suggest that the frequency of fire can decrease despite warmer temperatures because of increased precipitation. The warmer and drier CGCM1 produces a 30% increase in SSR for the southeastern United States and Alaska, with about 10% increases elsewhere. These scenarios suggest an increase in fire intensity and a 25%-50% increase in the area burned in the United States. In addition, recent results from the MC1 model, which is described by Neilson and Drapek, show an increase in area and biomass burned under both scenarios. This model includes an interaction with CO_2 concentrations, which, through increased CO., fertilization and increased water-use efficiency, produces more biomass and thus more fuel, contributing to more and larger fires under a highly variable climate that has dry years interspersed with wet periods.

Drought

Droughts occur in nearly all forest ecosystems. Drought effects are influenced by soil texture and depth; exposure; species present; life stage; and the frequency duration, and severity of drought. Droughts occur irregularly in forests of the humid regions east of the Mississippi River and in the superhumid Pacific Northwest. Droughts occur annually at the end of the growing season in forests at the midcontinen-tal prairie-forest border, where annual precipitation ranges from 600-1000 mm, or within humid regions that have shallow or rocky soils. Seasonal summer droughts are experienced by western interior dry forests that depend on winter precipitation, such as forests in the semiarid plains and in-termountain regions of the western United States. In some regions, droughts last several years.

The primary immediate response of forests to drought is to reduce net primary production (NPP) and water use, which are both driven by reduced soil moisture and stomatal conductance. Under severe conditions, plants die. Small plants, such as seedlings and saplings, are usually the first to die and can succumb under moderate conditions. Deep rooting and stored carbohydrates and nutrients make large trees susceptible only to severe droughts. Secondary effects also occur. When reductions in NPP are extreme or sustained over multiple growing seasons, increased susceptibility to insects or disease is possible, especially in dense stands. Drought can also reduce decomposition processes, leading to a buildup of organic matter on the forest floor that may increase fire frequency or intensity or reduce nutrient cycling.

The consequences of drought depend on annual and seasonal climate changes and on whether the current drought adaptations are sufficient to confer resilience to new conditions. Forests tend to grow to a level of maximum leaf area that nearly fully uses soil water during the growing season. A small increase in growing-season temperature could increase evaporative demand, triggering moisture stress. New results from two models described by Daly et al, MAPSS and MC1, suggest that this mechanism may cause future increases in drought stress in the Southeast, southern Rockies, and parts of the Northwest. The MC1 model indicates that the Prairie Peninsula and Great Lakes region, parts of the Northwest, and the Gulf Coast could experience drought stress within two decades, even though these regions may become wetter in later decades.

Insect and Pathogen Outbreaks

Climate influences the survival and spread of insects and pathogens directly, as well as the susceptibility of their forest ecosystems. Changes in temperature and precipitation affect herbivore and pathogen survival, reproduction, dispersal, and distribution. Indirect consequences of disturbance from herbivores and pathogens include elimination of nesting trees for birds and negative effects on mycorrhizal fungi. Other indirect effects include the impacts of climate on competitors and natural enemies that regulate the abundance of potential pests and pathogens.

Changes in the intensity and frequency of herbivore and pathogen damage in forests can have a range of effects. Most tree species support a community of other organisms, so the loss of any tree species can significantly reduce overall biodiversity. Such a loss occurred when chestnut blight almost completely eliminated chestnut trees ; the die-off of Fraser fir *(Abies fraseri)* caused by balsam woolly adel-gid *(Adelges piceae)* also raises concerns. Herbivore and pathogen damage to trees can increase understory plant diversity ; the overall abundance and diversity of birds ; and the diversity of predators, parasitoids, and detritivores.

Because climate change can both directly and indirectly affect herbivores and pathogens through various processes, the ultimate effects on patterns of disturbance include increased disturbance in some areas and decreased disturbance in others. For example, an increase in the interannual variation in minimum winter temperatures is expected to favor more northerly outbreaks of southern pine beetles but could reduce more southerly outbreaks. Similarly, decreased precipitation and increased evapotranspiration should boost tree secondary chemical metabolism (and, therefore, resistance to pests) in forests that currently suffer modest growing-season water deficits.

If global warming shifts species abundances, there may be associated shifts in herbivory Compared to the cooler Paleo-cene, the Eocene had a greater diversity of herbivores and higher attack rates on the most abundant tree species. Increased warming would most likely increase the diversity of insects at higher latitudes. Because insects typically migrate much faster than trees, many temperate tree species are likely to encounter nonnative insect herbivores that previously were restricted to subtropical forests.

Introduced Species

Introduced species can affect forests through herbivory, predation, habitat change, competition, alteration of gene pools via hybridization with natives, and disease (as either pathogens or vectors). Introduced species can alter the diversity, nutrient cycles, forest succession, and fire frequency and intensity of some ecosystems. The effects of introduced species should be considered concurrently with changes in native species distribution and abundance that occur as a consequence of climate change. The impact of introduced species on ecosystems is influenced by such climatic factors as temperature, drought, and cloud cover. Invasion biology is not yet adept at forecasting impacts of invasions. The complex interactions among introduced species, native communities, managed and intensely harvested forests, and climate change compound this forecasting problem.

The ultimate ranges of introduced species are largely determined by climate and human activities. Climate change will modify the distributions of many introduced species. Developmental rates will be modified by temperature change. For example, laboratory studies of balsam woolly adelgid growing under various temperature conditions provided the basis for simulations that suggest that temperature-induced changes in the population dynamics of the insect significantly affect Fraser fir survival.

The great majority of introduced species do not survive. Many fail because the climate is unsuitable at their points of arrival. Thus, a changed climate will lead to a different mix of surviving and failing species. In general, one might expect a larger fraction of survivors when the climate is warmer; introduced species comprise a far larger fraction of the biota in the warmer areas of the United States.

Increased CO., can directly influence introduced plants through enhanced photosynthesis, but at different rates for different species. Resistance of trees to introduced herbivores is sensitive to both climate and CO_2 concentrations. Climate change, in concert with CO2 concentration and nitrogen de-position, affects leaf nitrogen, which in turn influences herbivory

Hurricanes

Hurricanes disturb forests of the eastern and southern coastlines of the United States, as well as those of the Caribbean islands and the Atlantic coast of Central America. Ocean temperatures and regional climate events influence the tracks, size, frequency, and intensity of hurricanes. An average of two hurricanes make land every 3 years in the United States. Global warming may accelerate the hydrologic cycle by evaporating more water, transporting that water vapor to higher latitudes, and producing more intense and possibly more frequent storms. However, other variations may override possible increases in hurricane frequency.

Changes in the global hydrologic cycle and temperature will influence hurricane formation, but we cannot yet predict the direction and magnitude of change. Sea-surface temperatures are expected to rise, with hotter temperatures expanding to higher latitudes. Most studies point to an increase in hurricane frequency. However, even if frequency does not increase, it is likely that intensity and possibly duration of individual storms will increase because of the warming of the air and ocean, sources of energy for a hurricane.

The effects of hurricanes on vegetation include sudden and massive tree mortality, complex patterns of tree mortality (including delayed mortality), and altered patterns of forest regeneration. These changes can lead to shifts in successional direction, higher rates of species turnover, and opportunities for species change in forests, which can in turn increase landscape heterogeneity, produce faster biomass and nutrient turnover, and result in lower aboveground biomass in mature vegetation. Hurricanes can also result in buried vegetation and carbon sinks.

Windstorms

Small-scale wind events are products of mesoscale climatic circumstances and thus may be affected by climate changes, although the type and amount of alteration in windstorm characteristics cannot be predicted because these smaller-scale events are below the resolution of today's GCMs. Yet, tornadoes, downbursts, and derechos (a series of storm cells along a squall line) are probably the most important agents of abiotic disturbance to eastern deciduous forests. These disturbances can create very large patches of damage: A windstorm on 4 July 1999 in the Boundary Waters Canoe Area of Minnesota flattened roughly 250,000 acres of forest. Windstorms can cause heavy mortality, produce canopy disruption, reduce tree density and size structure, and change local environmental conditions. Consequently, the disturbance may trigger advance regeneration, seed germination, and accelerated seedling growth. These effects can change successional patterns, gap dynamics, and other ecosystem-level processes. The relationship between wind strength and severity of disturbance is not constant across different forests and species; although shallow-rooted species and thinned stands may be especially vulnerable to wind events, multiple factors influence tree response to high winds.

Berz suggests that increased intensity of all atmospheric convective processes will accelerate the frequency and intensity of tornadoes and hailstorms. Consistent with this view, Karl and colleagues found that the proportion of precipitation occurring in extreme thunderstorm events increased in the United States from 1910 to 1990, and Karl and colleagues further suggest that the climate of the United States has become more extreme (in terms of temperature and precipitation anomalies) in recent decades. Thus, it appears that the thunderstorm conditions that contribute to tornado formation have increased and are likely to continue increasing under projected climate changes. Furthermore, Etkin found a positive correlation between monthly tornado frequency and mean monthly temperature in western Canada, and inferred that this relationship suggests increased tornado frequency under a warmer climate scenario. Despite these inferences about tornado frequencies and the direct data on thunderstorm trends, understanding of tornado genesis is still inadequate to allow a direct forecast of how climate change will affect the frequency or severity of windstorms in the next century.

Ice Storms

Ice storms are caused by rain falling through sub-freezing air masses close to the ground; those air masses supercool the raindrops, which freeze on impact. Ice accumulation can vary dramatically with topography, elevation, exposure, and areal extent of the region over which conditions favor glaze formation. Ice storms occur throughout the United States except along the southwestern borders and parts of the plains, but the frequency and severity of ice storm events increase toward the northeastern US borders. However, the historic record of ice-storm events over large areas has not been consistent or precise, with rigorous measurements of ice accumulation.

Ice storms affect trees, forests, and forested landscapes in different ways. Ice damage to trees can range from severing a few twigs, to bending stems, to moderate crown loss, to outright breakage of trunks. Depending on stand composition, amount and extent of ice accumulation, and stand history, damage to stands can range from light and patchy to total breakage of all mature stems. Effects on forest stands include shifts in overstory composition in favor of more resistant tree species, loss of stand growth until leaf area is restored, and damage to stem form. Damaged stems are then more susceptible to the impacts of insects and disease. Recently thinned stands can be highly vulnerable because crowns have spread into the new space but branch strength has not developed. Several tree species can survive within areas frequented by ice storms. Though weather conditions producing ice storms are well understood, it is unclear how changes in climate will affect their frequency, intensity, regional location, or areal extent. However, atmospheric warming will most likely shift the locations of prevailing ice storms northward.

Landslides

Both slow and rapid movements of soil, rock, and associated vegetation are triggered directly by climate factors and indirectly by climate-influenced processes and by nonclimate factors such as earthquakes and volcanism. Triggering climatic events include snowmelt and intense rainfall, including that associated with hurricanes. Landslide frequency and extent are influenced by precipitation amount and intensity; snow accumulation, melt rate, and distribution; and roads and other land uses. The potential for a site to slide is influenced by slope steepness, properties of soil and rock, and hydrologic factors. Vegetation influences the likelihood of sliding through the soil-stabilizing effects of root systems and the effects of vegetation structure and composition on hydrology. Landslides remove soil and vegetation from steep slopes and damage forests on gentler slopes where landslide deposits come to rest. Landslides in forest landscapes can also damage aquatic resources and threaten public safety. Yet it is important to recognize that landslides are natural components of terrestrial and aquatic ecosystems.

Climate-change effects on landslides reflect changes in the delivery of water to soils through altered precipitation and snow hydrology. Because climate change is expected to vary geographically and with elevation, landslide responses will vary with similar complexity. Landslides are expected to be less frequent in areas where GCM scenarios predict reduced overall precipitation or reduced snowmelt because of warming trends, limiting snow accumulation. In the Pacific Northwest, much of the small, rapid landsliding occurs during rain-on-snow events in a broad elevation band where snow accumulates and melts several times in an average year. A simple warming without change in overall annual precipitation would be expected to result in reduced sliding by limiting the amount of snow (and its associated snowmelt) available to augment the rainfall reaching the soil. The most socially and ecologically significant landslides are triggered by intense precipitation. Thus climate change that increases storminess, and hence soil saturation, will increase landslide occurrence.

Interactions among Disturbances

Many disturbances are cascading. Drought often weakens tree vigor, leading to insect infestations, disease, or fire. Insect infestations and disease promote future fires by increasing fuel loads, and fires promote future infestations by compromising tree defenses. Increased fire intensity or extent would enhance the potential for landslides. Also, changes in land use, forest management, and atmospheric chemistry can interact with these natural disturbances. For example, harvest and road establishment in landslide-prone areas coupled with increased wetness could result in more landslides. In the southern Appalachians, ozone exposure coupled with infestations of exotic insects and climate change may increase Fraser fir mortality and red spruce stress. In some cases, however, the combination of disturbances may ameliorate impacts. Under droughty conditions, stomata tend to close, reducing the effects of high ozone exposure.

Nevertheless, when ecosystems experience more than one disturbance, the compounded effects can lead to new domains or surprises. A new domain is entered when the system has not recovered from the first disturbance before a second perturbation occurs, leading the system to a new long-term condition. For instance, the combination of climatically driven wildfires, fragmentation caused by agricultural settlement, and logging in the boreal forest has resulted in significant and unprecedented changes in forest composition. Invasive nonnative species are sometimes able to modify existing disturbance regimes or introduce entirely new disturbances. Under climate change, these compounded interactions may be unprecedented and unpredictable. They are likely to appear slowly and be difficult to detect because trees live for so long.

Strategies for Dealing with Forest Disturbances

Coping strategies for forests are influenced by the value of the forest, the naturalness of the disturbance, and the range of acceptable management options. Often the least ecologically disruptive response after a disturbance is no action at all, but managers or society usually call for some type of cleanup or restoration, even when such action may retard recovery. The value and management goals for the forest dictate how many resources can be allocated to its management. These values can change, as is illustrated by the revision of burn policy to recognize fire as a natural part of forest development that should not always be controlled.

The ability to manage for these eight disturbances varies greatly. However, current understanding of the disturbance nearly always provides some guidance for management under a future changed climate. Coping strategies for one disturbance type are often appropriate management responses to other disturbance types. For example, the removal of dead or dying trees and downed woody debris can reduce the risk of fire as well as alter insect and disease dynamics. Density management can reduce drought stress as well as alter insect population dynamics, but it could make forests more susceptible to wind. Thus, management effects are not always positive. Strategies for coping with disturbances in forests may also vary regionally. No matter where they are carried out, however, these practices often take 50-100 years to convert a landscape, and they are difficult to implement on inaccessible sites or in reserves.

We organize the coping strategies into several categories: managing before the disturbance, managing the disturbance itself, managing the recovery, and monitoring for adaptive management. These options are presented independent of climate-change effects but with the understanding that climate may alter the disturbance regime.

Managing before a Disturbance

Before a disturbance occurs, forests can be managed to reduce vulnerability or to enhance recovery. In both cases, management actions can alter the structure or the composition of the forest. In situations where the goal is to reduce the chance of future disturbances, adjustments to forest structure can be useful. For example, species or individual trees susceptible to ice or wind storms can be removed, as is common in cities. In addition, tree spacing and density can be altered to reduce susceptibility to drought. However, dead woody debris has numerous benefits, and its extensive removal can affect the biota and nutrient cycling. Managers can also change species composition to reduce the vulnerability of forests to disturbances. Tree species that are less vulnerable to fire, droughts, wind, insects, or pathogens can be planted or maintained. For example, the colonization of phloem-feeding insects, such as bark beetles, is partially controlled by the ability of the tree to produce oleoresin, which is under genetic control. So, planting selected tree species and genotypes with relatively high oleoresin could limit insect outbreaks.

Landscape structural changes can also reduce the chances that future disturbances will damage the forest. The pattern of clear-cutting influences the potential for windstorms to blow down trees, because destructive winds are more prevalent along the edge of a cut. And the placement of roads can influence the likelihood of future landslides and the spread of wildfire.

Management can be designed to reduce the opportunity for disturbance to occur. Examples are regulations that limit the introductions of nonnative species, the imposition of burning restrictions, and the use of controlled burns to reduce fuel loads. Trees can be planted that are less susceptible to disturbance. Species that promote disturbances can be removed. Density of trees can be managed to reduce the potential for future insect outbreaks or storm damage. Finally, roads can be designed to reduce the potential for landslides.

Other management actions can enhance forest recovery. Forest structure can be modified to speed up the succes-sional process in the event of a disturbance. Alternatively, species composition can be adjusted to promote recovery. For example, in areas likely to experience a disturbance, trees with salvage value can be planted.

Managing the Disturbance

Some disturbances, such as fire, insects, disease, and drought, can be managed during the disturbance through preventive measures or manipulations that affect the intensity or frequency of the disturbance. Alternatively, the disturbance can be managed to reduce its impact. A common way to control outbreaks of the southern bark beetle is to be on the alert for sites experiencing some beetle damage, then to cut those trees quickly to reduce the size of the area affected. Fire control is another example of a management action to reduce the impact of a disturbance.

Managing the Recovery

Recovery efforts can focus either on managing the state of the system immediately after the disturbance (e.g., salvage logging) or managing the ongoing process of recovery (e.g., planting and reseeding). Recovery efforts need careful consideration of the long-term impacts because such actions can damage soils and residual trees. Stands can recover naturally without any removal of the dead or damaged trees.

Recovery actions can be designed to speed recovery. In the aftermath of a disturbance, recovery can be enhanced by adding structural elements that create shade or other safe sites necessary for reestablishing vegetation or that serve as perches for birds (and thus places where seeds would be dispersed). Alternatively, late successional species can be planted to speed up succession. Finally, additions of water or nutrients can reduce environmental stress and facilitate restoration. Recovery can also be managed to reduce vulnerability to future disturbances.

Monitoring for Adaptive Management

A monitoring program should be used to determine how disturbances affect forests and to continually update our understanding of how climate change is potentially influencing the disturbance regimes. Monitoring can be designed to measure the state of the forest with and without disturbance under different management activities or to identify potential risks of forest disturbances. Such information is used to inform management of the potential outcomes of management actions.

Although many coping strategies associated with these disturbances could be incorporated into current forest-management practices regardless of climate change, the potential changes in climate may create a novel disturbance. For example, climate change may allow the migration of non-indigenous species into a forest, and current understanding of interactions and coping strategies may not apply to the resulting competitive interactions between nonindigenous and native species. Adaptive management approaches management as a continual learning process. The continued monitoring of ecosystem structure and function could be part of the coping strategy to address the likely surprises. The impacts of insects and pathogens are already monitored through the Forest Health Program, and weather and fuel moisture are monitored to assess the risk of fire during the fire season. However, few surveys quantify the extent and severity of damage from wind and ice storms or landslides.

Information from monitoring programs could be used to update risk assessments in management plans and prescriptions in an adaptive-management sense. A risk-ranking system could identify aspects of the forest most susceptible to disturbance under a changing climate. In conjunction with spatially explicit modeling of the site under various scenarios of disturbance impacts, a risk map could be created to identify sites most in jeopardy.

Research Needs

A key feature of this analysis is the realization of our lack of knowledge in many critical areas. We determined the key research needs for each disturbance and then organized the questions that must be resolved into the six topics discussed below.

Understanding Climatological Conditions that Initiate Disturbance

Accurate projections of climate effects of disturbances require improved climate and weather forecasts. The projections should include not only average climate conditions but also their range and variance. Short-term weather forecasts will be needed to predict drought occurrences for existing forests. For long-term climate change projections, improved resolution in climate models is needed so that regional patterns can be projected.

We have limited understanding of what climatological conditions lead to some disturbances. Improved understanding of local meteorological events that spawn tornadoes is needed, as well as improved projections of conditions that foster thunderstorms. Our ability to predict the occurrence of fires and hurricanes has benefited from research that allows managers to focus their attention on sites most likely to be disturbed. However, some disturbances result from interactions between ecological and climatological conditions that are often poorly understood. For example, better monitoring is needed to improve the characterization of ice accumulation in relation to storm characteristics and associated weather, especially the delineation of areas by amount of ice accumulation. Once the relationship between climate and disturbances has been quantified, more-accurate predictions of disturbances can be developed to minimize their impact.

Understanding the Effects of Disturbances on Microclimate

Because land-cover patterns can affect atmospheric circulation and cloud formation, changes in forest structure in the aftermath of fire, wind or ice storms, hurricanes, landslides, drought, and pest outbreaks may alter weather or climate conditions. This interaction needs to be studied and better understood.

Quantifying Impacts of Disturbances on Forests

There is a paucity of basic information on the frequency, intensity, and spatial extent of some disturbances and their impacts on forests. This problem is especially severe for landslides, ice storms, and small wind events. For example, reconstructive studies should be done to determine the long-term influence of successive ice storms on forests. Such analysis also allows exploration of interactions between disturbances and delayed responses.

Research should identify herbivores and pathogens that are likely to be key agents of forest disturbance in the next 50 years. Integrated continental surveys are needed to determine the sensitivity of different types of pests and diseases to environmental change and the potential for increased outbreaks of insect herbivores and pathogens at the margins of their existing ranges.

Interactions between Forest Disturbances and Management

Our ability to manage forests now as well as under climate change rests on our understanding of how forests respond to multiple disturbance events. A better understanding of interactions among fire, hurricanes, and biological disturbances (such as insects, pathogens, and introduced species) would improve our long-range predictions about forest succession and ecosystem dynamics and would lead to better prediction of conditions under which one event would affect the response to a subsequent one. This understanding, however, is complicated by the diverse goals of forest management.

Some management practices have been developed to cope with the physical disturbances of droughts, hurricanes, and wind events. However, additional research could expand options for management. Research is needed on the mitigation of hurricane impacts (i.e., how to hurricane-proof landscapes and how to design protected areas, for example, determining what their area, shape, and distribution should be). Forest ecologists and land managers are exploring the prospects for tailoring forest management regimes to the range of ecosystem conditions and wildfire disturbance regimes observed in the past, and in some cases to those anticipated under future climate conditions. For drought, new field experiments could test forest sensitivity to specific climate-change projections in combination with changes in the concentration of atmospheric trace gases. How the genetic diversity of host plants will determine the future epidemiology of forest pathogens needs further exploration. Critical evaluations of known patterns of species change and yield following past climate changes are needed, along with models of succession that incorporate disturbance processes.

Forest and Ecosystem Management in the Face of Change

During the last several decades, forest managers have relied on paradigms of ecological sustainability, historical variability, and ecological integrity to set goals and inform management decisions. These concepts commonly use historical forest conditions, usually defined as those that occurred before Euro-Americans dominated North American landscapes, as a means of gaining information about how healthy forests should be structured. There is no doubt that historical data have immense value in improving our understanding of ecosystem responses to environmental changes and setting management goals. However, many forest managers also use the range of historical ecosystem conditions as a management target, assuming that by restoring and maintaining historical conditions they are maximizing chances of maintaining ecosystems (their goods, services, amenity values, and biodiversity) sustainably into the future. This approach is often taken even as ongoing climate changes push global and regional climates beyond the bounds of the last several centuries to millenia. As importantly, novel anthropogenic stressors such as pollution, habitat fragmentation, land-use changes, invasive plants, animals, and pathogens, and altered fire regimes interact with climate change at local to global scales. The earth has entered an era of rapid environmental changes that has resulted in conditions without precedent in the past no matter how distantly we look. Attempts to maintain or restore past conditions require increasingly greater inputs of energy from managers and could create forests that are ill adapted to current conditions and more susceptible to undesirable changes. Accepting that the future will be different from both the past and the present forces us to manage forests in new ways. Further, although quantitative models can estimate a range of potential directions and magnitudes of environmental changes and forest responses in the future, models rarely can predict the future with the level of accuracy and precision needed by resource managers. We might feel confident of broad-scale future environmental changes (such as global mean temperature increases), but we cannot routinely predict even the direction of change at local and regional scales (such as increasing or decreasing precipitation). A healthy skepticism leads us to use models to help organize our thinking, game different scenarios, and gain qualitative insight on the range of magnitudes and direction of possible future changes without committing to them as forecasts.

Facing an unknowable and uncertain future, however, does not mean "anything goes" for natural resource management. Managing in the face of uncertainty will require a portfolio of approaches, including short-term and long-term strategies, that focus on enhancing ecosystem resistance and resilience as well as assisting forested ecosystems to adapt to the inevitable changes as climates and environments continue to shift. Historical ecology becomes ever more important for informing us about environmental dynamics and ecosystem response to change. We offer here a conceptual framework for developing forest management strategies in a context of change.

The premise of an uncertain but certainly variable future is effectively best addressed with approaches that embrace strategic flexibility, characterized by risk-taking (including decisions of no action), capacity to reassess conditions frequently, and willingness to change course as conditions change. Learning from experience and iteratively incorporating lessons into future plans (adaptive management in its broadest sense) is the necessary lens through which natural resource management must be conducted. Decisions that emphasize ecological process, rather than structure and composition, become critical. An example is increased use of managed wildfire in remote places. Similarly, institutional flexibility will be more effective than rigid or highly structured decision making.

A central dictum under uncertain futures is that no single approach will fit all situations. A toolbox approach, from which various treatments and practices can be selected and combined to fit unique situations, will be most useful. Some applications will involve traditional management approaches, but used in new locations, seasons, or contexts. Other options may require experimenting with new practices. A toolbox approach recognizes that strategies may vary based on the spatial and temporal scales of decision-making. Planning at regional scales will often involve acceptance of different levels of uncertainty and risk than appropriate at local scales.

The framework of options presented below includes both adaptation strategies, that is, actions that help forested ecosystems accommodate changes, and mitigation strategies, actions that reduce the causes of stress, such as reducing anthropogenic climate change by sequestering CO_2 and reducing greenhouse gases. Integrative approaches that combine adaptation and mitigation practices in complementary ways are favored. A first consideration in building an integrative strategy is to evaluate the types of uncertainty. These could include, for example, knowledge about present environmental and ecological conditions, models and information sources about the future, institutional resources (staff, time, funds available), planning horizon (short- vs. long-term), and public and societal support A further decision is whether, or to what degree, to adopt deterministic or indeterministic approaches. The former accepts certain kinds of information about the future as reliable enough upon which to base decisions. By contrast, indeterministic approaches base planning on an assumption that information about the future is not adequately known, and plan instead directly for uncertainty.

Deterministic approaches "put all the eggs in one basket" and risk potential failures if an assumed future does not unfold, whereas indeterministic approaches employ "bet hedging" strategies that attempt to minimize risks by taking multiple courses of action. Below we offer management options and examples for populating a manager's climate-change toolbox.

Create Resistance to Change

One set of adaptive options is to manage forest ecosystems and resources so that they are better able to resist the influence of climate change or to forestall undesired effects of change. Whereas this may seem a denial of future change, it is a defensible approach to uncertainty. From high-value plantations near harvest to high-priority endangered species with limited available habitat, maintaining the status quo for a short time may be the only or best option. Resistance practices seek to improve forest defenses against direct and indirect effects of rapid environmental changes. In western North America these will commonly include reducing undesirable or extreme effects of fires, insects, and diseases. Treatments might include complete fuel breaks around highest risk or highest value areas (such as wildland–urban interfaces, forests with high amenity or commodity values, or at-risk species); intensive removal of invasives; or interventions such as those used in high-value agricultural situations. Abrupt invasions, changes in population dynamics, and long-distance movements of native and nonnative species are expected in response to changing climates. Climate changes may also catalyze conversion of native insects or disease species into invasive species in new environments, such as with mountain pine beetle (Dendroctonus ponderosae) east of the Continental Divide in Canada. Taking early defensive actions at key migration points to remove and block invasions is important to increase resistance.

Resisting climatic and other environmental changes to forests often may require intensive intervention, accelerating efforts and investments over time, and a recognition that eventually these efforts may fail as conditions change cumulatively. Creating resistance to directional change is akin to "paddling upstream," and eventually conditions may change so much that resistance is no longer possible. For instance, site capacities may shift from favoring one species to another. Forests that have been treated to resist climate-related changes may cross thresholds and be lost catastrophically. For this reason, resistance options are best applied in the short-term and to forests of high value. Forests with low sensitivity to climate may be those most likely to accommodate resistance treatments, and high-sensitivity forests may require the most intensive efforts to maintain.

Promote Resilience to Change

Resilient forests are those that not only accommodate gradual changes related to climate but tend to return toward a prior condition after disturbance either naturally or with management assistance. Promoting resilience is the most commonly suggested adaptive option discussed in a climate-change context, but like resistance, is not a panacea. Resilience in forest ecosystems can be increased through practices similar to those described for resisting change but applied more broadly, and specifically aimed at coping with disturbance. Given that the plant establishment phases tend to be most sensitive to climate-induced changes in site potential, surplus seed-banking, and intensive management during revegetation through early years of establishment may enable retention of desired species, even if the site is no longer optimal.

Capacity to maintain and improve resilience may become more difficult and require more intensive intervention as changes in climate accumulate over time. These options are best exercised in projects that are short-term, have high amenity or commodity values, or under ecosystem conditions that are relatively insensitive to climate change effects.

Enable Forests to Respond to Change

This group of adaptation options intentionally accommodates change rather than resists it, with a goal of enabling or facilitating forest ecosystems to respond adaptively as environmental changes accrue. Treatments implemented would mimic, assist, or enable ongoing natural adaptive processes such as species dispersal and migration, population mortality and colonization, changes in species' dominances and community composition, and changing disturbance regimes. The strategic goal is to encourage gradual adaptation and transition to inevitable change, and thereby to avoid rapid threshold or catastrophic conversion that may occur otherwise.

Depending on the context, management goals, and availability and adequacy of modeling information (climate and otherwise), different approaches may be chosen. Changes in fundamental ecosystem state are assumed to happen, either in some general direction (deterministic) where specific goals are planned for the future, or in unknown directions (indeterministic) where goals are developed for uncertainty. A sample of potential practices follows.

1. *Assist transitions, population adjustments, range shifts, and other natural adaptations.*—Qualitative indications of future change may be adequate to trigger actions at least in broad outline. With such

information, managers might plan for transitions to new conditions and habitats, and assist the transition, e.g., as appropriate, assist species migrations along expected climatic gradients, plan for higher-elevation insect and disease outbreaks, anticipate forest mortality events and altered fire regimes, or accommodate loss of species' populations on warm range margins (Ledig and Kitzmiller 1992, Parker et al. 2000). For forest plantations, examples would include modifying harvest schedules, altering thinning prescriptions and other silvicultural treatments, replanting with different species, shifting desired species to new plantation or forest locations, and taking precautions to mitigate likely increases in stress on plantation and forest trees.

A nascent literature explores the advantages and disadvantages of "assisted migration," that is, intentional movement of propagules or juvenile and adult individuals into areas assumed to be their future habitats. Some environments have broad and regular gradients, making adaptive migration directions obvious. Others, such as patchy mountainous terrain, are heterogeneous, and migration direction is far more difficult to determine. On-the-ground monitoring of native species can provide insight into what organisms are experiencing, and indicate the directions of change and appropriate response at local scales. This can allow management strategies to mimic emerging natural adaptive responses rather than rely on quantitative projections. For instance, new species mixes (mimicking what is regenerating naturally or outperforming plantation species), altered genotype selections, modified age structures, and new management contexts (e.g., uneven vs. even-aged management, altered prescribed fire regimes) may be considered.

2. *Increase redundancy and buffers.*—Here we suggest using redundancy and creating diversity through practices that spread risks rather than concentrate them.These can be achieved, for instance, by introducing species over a range of environments rather than withinhistorical distribution, "preferred habitat," or projectedfuture environments. Redundant plantings across arange of environments can provide monitoring information if survival and performance are measured and analyzed. Reexamining replicated forest plantations, such old genetic provenance or progeny tests, is a means of gathering information about adaptation to recent and ongoing changes. Opportunistic assessment, such as of horticultural plantings of native species in landscaping, gardens, roadsides, or parks, can give clues on how species respond

in different locations as climate changes.

3. *Expand genetic diversity guidelines.*—Existing guidelines for genetic management of forests and restoration projects specify actions to retain local gene pools. In the past, strict transfer rules that minimized movement of germplasm and small seed zones were developed to avoid contamination of populations with ill-adapted genotypes. These rules were based on assumptions that neither environments nor climate were changing. Relaxing these guidelines may be appropriate under assumptions of changing climates. In this case, either deterministic or indeterministic options could be chosen. In the former, germplasm would be moved in the expected adaptive direction, for instance, rather than using local seed, seed from a warmer population would be used. New transfer rules could be developed for expected future climate gradients. By contrast, if an uncertain future is assumed, expanding seed zone sizes or relaxing rules to admix germplasm from adjacent zones might be considered. Adaptive management of this nature is experimental by design, should be undertaken cautiously, and requires careful documentation of treatments, seed sources, and outplanting locations to learn from both failures and successes.

   Enforcing traditional best genetic management practices that equalize germplasm contributions and enhance effective population sizes becomes especially important under uncertain futures. Genotypes known or selected for broad adaptations would also be favored. By contrast, using a single or few genotypes (e.g., a select clone or small clonal mix) is far riskier in a long-term context of uncertainty.

4. *Manage for asynchrony and use establishment phase to reset succession.*—Changing climates over paleohis- torical time scales have repeatedly altered biotic communities as plants and animals responded to natural changes. To the extent that climate acts as a region- and hemispheric-wide driver of change, the resulting shifts in biota often occur as synchronous changes across the landscape. At decadal and centennial scales, for instance, recurring droughts in the west and windstorms in the east have synchronized forest composition and age- and stand structure across broad landscapes, which then become vulnerable to climate shifts. This appears to have happened in some western forests as widespread drought has induced diebacks. Opportunities exist to manage early successional stages following

widespread mortality by deliberately reducing landscape synchrony . Asynchrony can be achieved by promoting diverse age classes, species mixes, within-stand and across-landscape structural diversities, and genetic diversity. Early suc-cessional stages provide the most practical opportunities for resetting ecological trajectories in ways that are adaptive to present and future rather than past conditions.

5. *Establish ''neo-native'' forests.*—Information from historical species ranges and responses to climate change can provide unique insight about species responses, ecological tolerances, and potential new habitats. Areas that supported species in the past under similar conditions to those projected for the future might be considered sites for ''neo-native'' stands of the species. These may even be outside the current species range, in locations where the species would otherwise be considered exotic. For instance, Monterey pine (Pinus radiata), endangered throughout its small native range, has naturalized along the north coast of California distant from its present native distribution. Much of this area was paleohistorical range for the pine, extant during climate conditions that have been interpreted to be similar to expected futures in California. Using these locations for ''neo-native'' conservation stands, rather than removing trees as undesired invasives, is an example of how management could accommodate climate change.

6. *Promote connected landscapes.*—The capacity to move (migrate) in response to changing climates has been key to adaptation and long-term survival of plants and animals in historical ecosystems. Plants migrate (shift ranges) by dying in unfavorable sites and colonizing favorable sites, including internal species' margins. The capacity to do this is aided by managing for connected landscapes, that is, landscapes that contain continuous habitat with few physical or biotic impediments to migration, and through which species can move readily. Promoting connected forested landscapes with flexible management goals that can be modified as conditions change may assist species to respond naturally to changing climates. Desired goals include reducing fragmentation and planning at large landscape scales to maximize habitat connectivity.

7. *Realign significantly disrupted conditions.*—For forests that have been significantly disturbed and are far outside historical ranges of variation, restoration treatments are often prescribed. Re-alignment or en-

trainment with current and expected future conditions rather than restoration to historical pre-disturbance conditions may be a preferred choice In this case, management seeks to bring processes of the disturbed landscape into the range of current or expected future environments. The Mono Basin case in California exemplifies this approach, where water balance models were used to determine appropriate lake levels buffered for current and expected future climate variability.

8. *Anticipate surprises and threshold effects.*—Evidence is accumulating that species interactions and competitive responses under changing climates can be complex and unexpected. Managers can evaluate the potential for indirect and surprise effects that may result from cumulative climate changes or changes in extreme weather events. This involves anticipating events outside the range of conditions that have occurred in recent history. For example, reductions in mountain snowpacks lead to more bare ground in spring such that even ''average'' rain events may run off immediately, rather than being buffered by snowpacks, and produce extreme unseasonal floods. In many parts of western North America, additional stresses of extended summer water deficits are pushing plant populations over thresholds of mortality, as occurred in the recent multi-year droughts in the Southwest. Other examples already observed in some areas are year-round fire seasons and fires in atypical locations, such as subalpine and coastal environments.

9. *Experiment with refugia.*—Plant ecologists and paleoecologists recognize that some environments are more buffered against climate change and short-term disturbances than others. If such environments can be identified, they could be considered sites for long-term retention of plants or for establishment of new forests. For instance, microclimates in mountainous regions are highly heterogeneous. Furthermore, unusual and nutritionally extreme soil types (e.g., acid podsol, ultramafic, limestone) have been noted for their long persistence of species and genetic diversity, resistance to invasive species, and long-lasting community physiognomy compared to adjacent fertile soils. During historical periods of rapid climate change and widespread population extirpation, refugial populations have persisted on unusual local sites that avoided extremes of regional climate impacts or the effects of large disturbance.

Reduce Greenhouse Gases

This set of options has the goal of using forested environments to ameliorate greenhouse gas emissions and sequester carbon, thereby lessening the human impact on climate. The forestry sector has a huge potential to contribute at global to regional scales. Evaluating and determining best choices, however, are hampered by considerable uncertainty and difficulty in analyzing net carbon balances.

1. *Sequester carbon.*—Forest management strategies designed to achieve goals of removing CO_2 and storing carbon are diverse, and include avoiding deforestation, promoting afforestation and reforestation, manipulating vegetation to favor rapid growth and long-term site retention, and sequestering carbon after harvest in wood products. Some approaches duplicate long-recognized best forest-management practices, where goals are to maintain healthy vigorous trees, keep sites fully occupied with minimal spatial or temporal gaps in non-forest conditions, and minimize severe disturbance by fire, insects, and disease. As noted above, however, in many cases uniform forest conditions are best avoided, as they are vulnerable to mortality from insects, disease, and fire. Under changing climates, these conditions may need to be intensively managed to minimize risk of severe fire, and to reduce the potential for carbon losses from wildfire.

   Once wood is removed from the forest or plantation, its subsequent use affects its sequestration status. Options for minimizing return of carbon to the atmosphere include storing carbon in wood products, or using it as biomass to fuel electricity production, thereby providing alternative forms of energy to replace fossil fuels. For successful choices to be made, life-cycle analysis research must assess carbon accounting from forest through utilization phases.

2. *Reduce emissions.*—Wildfire and extensive forest mortality as a result of insect and disease are primary sources of unintentional carbon emissions from forests in western United States, and can lead to widespread loss of centuries' worth of carbon storage. This effect will likely be exacerbated in coming decades under continued warming, with increasingly severe fire years leading to what have been modeled as widespread ''brown-downs'' for many western and eastern forest types.

One obvious means of slowing this release of sequestered carbon is to increase forest resistance to fire, drought, and disease, usually by reducing the density of small trees. In roaded or otherwise accessible areas, such density reductions might be accomplished by mechanical thinning, prescribed fires, or both. In remote or rugged terrain, wildland fire use or appropriate management response suppression fire may be the only reasonable option. In either case, some carbon inevitably will be released in the process of increasing forest resistance to sudden release of much greater quantities of carbon. If small trees are physically removed during the density reduction, then subsequently used for energy generation or long-term sequestration, the net carbon release might be minimized.

Prioritizing Management Under Conditions of Rapid Change

Species respond to changing climates and environments individualistically. Some species will be sensitive and vulnerable whereas others will be naturally buffered and resilient to climate-influenced disturbances. Management goals across the spectrum of forest types and ownerships also vary. As a result, proactive climate planning will include a range of approaches having different management intensities. Some species and ecosystems may require aggressive treatment to maintain viability or resilience, others may require reduction of current stressors, and others less intensive management, at least in the near future.

Evaluating priorities has always been important in resource management. However, the magnitude and rate of change and the management responses these demand, combined with finite human resources and declining budgets, dictate that priorities be evaluated swiftly and definitively. A useful systematic approach for prioritizing high-demand situations might be adopted from the medical practice of triage. Deriving from the French word triare, to sort, triage approaches were developed from the need to prioritize care of injured soldiers in battlefield settings where time is short, needs are great, and capacity to respond is limited. Triage applied in a resource context offers a systematic process to sort management situations into categories according to urgency, sensitivity, and capacity of available resources to achieve desired goals. Cases are rapidly assessed and divided into three to five major categories that determine treatment priority. The categories range from high urgency (treat immediately), mid-urgency (treat later), to highly urgent but untreatable given current capacity (no action taken). Reassessing and re-prioritizing must be done frequently, especially when conditions are changing rapidly.

Although triage approaches are valuable under conditions of scarce resources or overwhelming choice, they are rarely adequate as long-term approaches. Other planning processes may be used for prioritizing current management plans and practices. An example is rapid assessments of forest management plans by teams of climate-expert reviewers who convene to intensively review existing management plans, assess current needs, and recommend top priorities for revision.

References

Aber J. (2001). Forest processes and global environmental change: The effects of individual and multiple stressors on forests. *BioScience* 51: 735–751.

Ayres MP, Lombardero MJ. (2000). Assessing the consequences of global change for forest disturbance from herbivores and pathogens. *Science of the Total Environment* 262: 263–286.

Dale VH, Joyce LA,McNulty S,Neilson RP. (2000). The interplay between climate change, forests, and disturbances. *Science of the Total Environment* 262: 201–204.

Flannigan MD, et al. (1998). Future wildfire in circumboreal forests in relation to global warming. *Journal of Vegetation Science* 9: 469–476.

Hansen AJ,Neilson RP, Dale VH, Flather CH, Iverson LR,Currie DJ, Shafer S, Cook R, Bartlein PJ. (2001). Global change in forests: Responses of species, communities, and biomes. *BioScience* 51: 765–779.

8

Impact of Climate Change on Water Resources

Since the beginnings of concern over the possible consequences of global warming, it has been widely recognized that changes in the cycling of water between land, sea, and air could have very significant impacts across many sectors of the economy, society, and the environment. The characteristics of many terrestrial ecosystems, for example, are heavily influenced by water availability and, in the case of instream ecosystems and wetlands, by the quantity and quality of water in rivers and aquifers. Water is fundamental to human life and many activities—most obviously agriculture but also industry, power generation, transportation, and waste management—and the availability of clean water often is a constraint on economic development. Consequently, there have been a great many studies into the potential effects of climate change on hydrology and water resources. The majority of these studies have concentrated on p o ssible changes in the water balance; they have looked, for example, at changes in streamflow through the year. A smaller number of studies have looked at the impacts of these changes for water resources—such as the reliability of a water supply reservoir or the risk of flooding—and even fewer explicitly have considered possible adaptation strategies. This assessment is based not only on the few studies that have looked explicitly at climate change but also on c o nsiderable experience within different parts of the water se ctor in adapting to changing circumstances in general.

Climate change is just one of many pressures facing the hydrological system and water resources. Changing land-use and land-management practices (such as the use of agrochemicals) are altering the hydrological system, often leading to deterioration in the resource baseline. Changing demands generally are increasing pressures on available resources, although per capita demand is falling in some countries. The objectives and procedures of water management are changing too: In many countries, there is an increasing move toward "sustainable" water management and increasing concern for the needs of the water environment. For example, the Dublin Statement, agreed at the International Conference on Water and the Environment in 1992, urg es su stainable use of water resources, aimed at ensuring that neither the quantity nor the quality of available resources are degraded. Key water resources stresses now and over the next few decades relate to access to safe drinking water, water for growing food, overexploitation of water resources and consequent environmental degradation, and deterioriation in water quality. The magnitude and significance of these stresses varies between countries. The late 1990s saw the development of several global initiatives to tackle water-related problems: The UN Commission on Sustainable Development published the "Comprehensive Assessment of the Freshwater Resources of the World", and the World Water Council asked the World Commission for Water to produce a vision for a "water-secure world". A series of periodical reports on global water issues was initiated. The impacts of climate change, and adaptation to climate change, must be considered in the context of these other pressures and changes in the water sector.

Estimating the Impacts of Climate Change

Over the past decade—and increasingly since the publication of the Second Assessment Report (SAR) —there have been many studies into climate change effects on hydrology and water resources, some coordinated into national programs of research (as in the U.S. National Assessment) and some undertaken on behalf of water management agencies. There are still many gaps and unknowns, however.

The impacts of climate change on hydrology usually are estimated by defining scenarios for changes in climatic inputs to a hydrologicalmodel from the output of general circulation models (GCMs). The three key developments here are constructing scenarios that are suitable for hydrological impact assessments, developing and using realistic hydrological models, and understanding better the linkages and feedbacks between climate and hydrological systems.

The heart of the scenario "problem" lies in the scale mismatch between global climate models (data generally provided on a monthly time step at a spatial resolution of several tens of thousands of square kilometers) and catchment hydrological models (which require data on at least daily scales and at a re solution of perhaps a few square kilometers). A variety of "downscaling" techniques have been developed and used in hydrological studies. These techniques range from simple interpolation of climate model output, through the use of empirical/statistical relationships between catchment and regional climate, to the use of nested regional climate models ; all, however, depend on the quality of simulation of the driving global model, and the relative costs and benefits of each approach have yet to be ascertained. Studies also have looked at techniques for generating stochastically climate data at the catchment scale. In principle, it is possible to explore the effects of changing temporal patterns with stochastic climate data, but in practice the credibility of such assessments will be strongly influenced by the ability of the stochastic model to simulate present temporal patterns realistically.

Considerable effort has been expended on developing improved hydrological models for estimating the effects of c l imate change. Improved models have been developed to s imulate water quantity and quality, with a focus on realistic representation of the physical processes involved. These models often have been developed to be of general applicability, with no locally calibrated parameters, and are increasingly using remotely sensed data as input. Although different hydrological models can give different values of streamflow for a given input, the greatest uncertainties in the effects of c l imate on streamflow arise from uncertainties in climate change scenarios, as long as a conceptually sound hydrological model is used. In estimating impacts on groundwater recharge, water quality, or flooding, however, translation of climate into response is less well understood, and additional uncertainty is introduced. In this area, there have been some reductions in uncertainty since the SAR as models have been improved and more studies conducted. The actual impacts on water resources—such as water supply, power generation, navigation, and so forth—depend not only on the estimated hydrological change but also on changes in demand for the resource and assumed responses of water resources managers. Since the SAR, there have been a few studies that have summarized potential response strategies and assessed how water managers might respond in practice.

There also have been considerable advances since the SAR in the understanding of relationships between hydrological processes at the land surface and processes within the atmosphere above. These advances have come about largely through major field measurement and modeling projects in different geographical environments, coordinated research programs (such as those through the International Geosphere-Biosphere Programme, Baltic Sea Experiment (BALTEX), and GEWEX Asian Monso. The ultimate aim of such studies often is to lead to improved assessments of the hydrological effects of climate change through the use of co upled climate-hydrology models; thus far, however, the be nefits to impact assessments have been indirect, through improvements to the parameterizations of climate models. A few studies have used coupled climate-hydrology models to forecast streamflow, and some have begun to use them to estimate effects of changing climate on streamflow.

Effect of Climatic Variability on Water Resources

Since the SAR, many studies have explored linkages between recognizable patterns of climatic variability—particularly El Niño and the North Atlantic Oscillation—and hydrological behavior, in an attempt to explain variations in hydrological characteristics over time. These studies in North America, South America, Australasia, Europe, and southern Afri ca have emphasized variability not just from year to year but also from decade to decade, although patterns of variability vary considerably from region to region. Most s t u dies focus on the past few decades with recorded h y d r o l o g ical data, but an increasing number of studies have reconstructed considerably longer records from various proxy data sources. Such research is extremely valuable because it helps in interpretation of observed hydrological changes over time (particularly at tr ibution of change to global warming), provides a context for assessment of future change, and opens up possibilities for seasonal flow prediction hence more efficient adaptation to climatic variability. It also emphasizes that the hydrological "baseline" cannot be assumed to be c onstant, even in the absence of climate change.

Adaptation to Climate Change in the Water Sector

Water management is based on minimization of risk and a da ptation to changing circumstances (usually taking the form of altered demands). A wide range of adaptation techniques has been developed and applied in the water sector over decades. One widely used classification distinguishes between increasing capacity (e.g., building reservoirs or structural flood defenses), changing operating rules for existing structures and systems, managing demand, and changing institutional practices. The first two often are termed "supply-side" strategies, whereas the latter two are "demand-side." Over the past few years, there has been a considerable increase in interest in demand-side techniques. International agencies such as the World Bank and initiatives such as the Global Water Partnership are promoting new ways of managing and pricing water resources to manage resources more effectively.

Water managers in some countries are beginning to consider climate change explicitly, although the methodologies for doing so are not yet well defined and vary between and within countries depending on the institutional arrangements for long-term water resources planning. In the UK, for example, water supply companies were required by regulators in 1997 to "consider" climate change in estimating their future resource, hence investment, projections. In the United States, the American Water Works Association urged water agencies to explore the vulnerability of their systems to plausible climate changes.

Effects on the Hydrological Cycle

Precipitation

Precipitation is the main driver of variability in the water b alance over space and time, and changes in precipitation have very important implications for hydrology and water resources. Hydrological variability over time in a catchment is influenced by variations in precipitation over daily, seasonal, annual, and decadal time scales. Flood frequency is affe cted by changes in the year-to-year variability in precipitation and by changes in short-term rainfall properties (such as storm rainfall intensity). The frequency of low or drought flows is affected primarily by changes in the seasonal distribution of precipitation, year-to-year variability, and the occurrence of prolonged droughts.

Current climate models simulate a climate change-induced increase in annual precipitation in high and mid-latitudes and most equatorial regions but a general decrease in the subtropics, although across large parts of the world the changes associated with global warming are small compared to those resulting from natural multi-decadal variability, even by the 2080s.

Changes in seasonal precipitation are even more spatially variable and depend on changes in the climatology of a region. In general, the largest percentage p recipitation changes *over land* are found in high latitudes, some equatorial regions, and southeast Asia, although there are large differences between climate models.

Until recently, very few projections of possible changes in year-to-year variability as simulated by climate models have been published, reflecting both the (until recently) short model runs available and the recognition that climate models do not necessarily reproduce observed patterns of climatic variability. Recent developments, however, include the increasing ability of some global climate models to reproduce features such as El Niño and open up the possibility that it may be feasible to estimate changes in ye a r-to-year variability. Recent scenarios for the UK, derived from HadCM2 experiments, indicate an increase in the relative variability of seasonal and annual rainfall totals resulting from global warming.

Potential changes in intense rainfall frequency are difficult to infer from global climate models, largely because of coarse spatial resolution. However, there are indications that the frequency of heavy rainfall events generally is likely to increase with global warming. Confidence in this assertion depends on the confidence with which global climate models are held. More generally, uncertainty in GCM projections of precipitation largely determines the uncertainty in estimated impacts on hydrological systems and water resources.

Increasing temperatures mean that a smaller proportion of precipitation may fall as snow. In areas where snowfall currently is marginal, snow may cease to occur—with consequent, very significant, implications for hydrological regimes. This projection is considerably less uncertain than possible changes in the magnitude of precipitation.

Evaporation

Evaporation from the land surface includes evaporation from open water, soil, shallow groundwater, and water stored on vegetation, along with transpiration through plants. The rate of evaporation from the land surface is driven essentially by meteorological controls, mediated by the characteristics of vegetation and soils, and constrained by the amount of water available. Climate change has the potential to affect all of these factors—in a combined way that is not yet clearly understood— with different components of evaporation affected differently.

The primary meteorological controls on evaporation from a well-watered surface (often known as potential evaporation) are the amount of energy available (characterized by net radiation), the moisture content of the air (humidity—a function of water vapor content and air temperature), and the rate of movement of air across the surface (a function of windspeed). Increasing temperature generally results in an increase in potential evaporation, largely because the water-holding capacity of air is increased. Changes in other meteorological controls may exaggerate or offset the rise in temperature, and it is possible that increased water vapor content and lower net radiation could lead to lower evaporative demands. The relative importance of different meteorological controls, however, varies geographically. In dry regions, for example, potential evaporation is driven by energy and is not constrained by atmospheric moisture contents, so changes in humidity are relatively unimportant. In humid regions, however, atmospheric moisture content is a major limitation to evaporation, so changes in humidity have a very large effect on the rate of evaporation.

Several studies have assessed the effect of changes in m e t e o rological controls on evaporation, using models of the evaporation process, and the effect of climate change has been shown to depend on baseline climate (and the relative importance of the different controls) and the amount of change. Chattopadhyary and Hulme calculated increases in potential evaporation across India from GCM simulations of climate; they found that projected increases in potential evaporation were related largely to increases in the vapor pressure deficit resulting from higher temperature. It is important to emphasize, however, that different evaporation calculation equations give different estimates of absolute evaporation rates and sensitivity to change. Therefore, it can be very difficult to compare results from different studies. Equations that do not consider explicitly *all* meteorological controls may give very misleading estimates of change.

Vegetation cover, type, and properties play a very important role in evaporation. Interception of precipitation is very much influenced by vegetation type (as indexed by the canopy storage capacity), and different vegetation types have different rates of transpiration. Moreover, different vegetation types produce different amounts of turblence above the canopy; the greater the turbulence, the greater the evaporation. Achange in catchment vegetation—directly or indirectly as a result of climate change—therefore may affect the catchment water balance (there is a huge hydrological literature on the effects of changing catchment vegetation). Several studies have assessed changes in biome type under climate change, but the hydrological effects of such changes—and, indeed, changes in agricultural land use—have not yet been explored.

Although transpiration from plants through their stomata is d r iven by energy, atmospheric moisture, and turbulence, plants exert a degree of control over transpiration, particularly when water is limiting. Stomatal conductance in many plants falls as the vapor pressure deficit close to the leaf increases, temperature rises, or less water becomes available to the roots—and t r a nspiration therefore falls. Superimposed on this short-term variation in stomatal conductance is the effect of atmospheric carbon dioxide (CO_2) concentrations. Increased CO_2 concentrations reduce stomatal conductance in C_3 plants (which include virtually all woody plants and temperate grasses and crops), although experimental studies show that the effects vary considerably between species and depend on nutrient and water status. Plant water-use efficiency (WUE, or water use per unit of biomass) therefore may increase substantially, implying a reduction in transpiration. However, higher CO_2 concentrations also may be associated with increased plant growth, compensating for increased WUE, and plants also may acclimatize to higher CO_2 concentrations. There have been considerably fewer studies into total plant water use than into stomatal conductance, and most empirical evidence to date is at the plant scale; it is difficult to generalize to the catchment or regional scale. Free-air CO_2 enrichment (FACE) experiments, however, have allowed extrapolation at least to the 20-m plot scale. Experiments with cotton, for example, showed no detectable change in water use per unit land area when CO_2 concentrations were increased to 550 ppmv; the 40% increase in biomass offset increased WUE. Experiments with wheat, however, indicated that increased growth did not offset increased WUE, and e vaporation declined by approximately 7%. Some model studies suggest that the net direct effect of increased CO_2 concentrations at the catchment scale will be small, but others indicate that stomata have more control on regional evaporation. There clearly is a large degree of uncertainty over the effects of CO_2 enrichment on catchment-scale evaporation, but it is apparent that reductions in stomatal conductance do not necessarily translate into reductions in catchment-scale evaporation.

The *actual* rate of evaporation is constrained by water availability. A reduction in summer soil water, for example, could lead to a reduction in the rate of evaporation from a catchment despite an increase in evaporative demands. Arnell estimated for a sample of UK catchments that the rate of actual evaporation would increase by a smaller percentage than the atmospheric demand for evaporation, with the greatest difference in the "driest" catchments, where water limitations are greatest.

Soil Moisture

The amount of water stored in the soil is fundamentally important to agriculture and is an influence on the rate of actual evaporation, groundwater recharge, and generation of runoff. Soil moisture contents are directly simulated by global climate models, albeit over a very coarse spatial resolution, and outputs from these models give an indication of possible directions of change. Gregory *et al.*, for example, show with the HadCM2 climate model that a rise in greenhouse gas (GHG) concentrations is associated with reduced soil moisture in Northern Hemisphere mid-latitude summers. This was the result of higher winter and spring evaporation, caused by higher temperatures and reduced snow cover, and lower rainfall inputs during summer.

The local effects of climate change on soil moisture, however, will vary not only with the degree of climate change but also with soil characteristics. The water-holding capacity of soil will affect possible changes in soil moisture deficits; the lower the capacity, the greater the sensitivity to climate change. Climate change also may affect soil characteristics, perhaps through changes in waterlogging or cracking, which in turn may affect soil moisture storage properties. Infiltration capacity and water-holding capacity of many soils are influenced by the frequency and intensity of freezing. Boix-Fayos *et al.*, for example, show that infiltration and water-holding capacity of soils on limestone are greater with increased frost activity and infer that increased temperatures could lead to increased surface or shallow runoff. Komescu *et al.* assess the implications of climate change for soil moisture availability in southeast Turkey, finding substantial reductions in availability during summer.

Groundwater Recharge and Resources

Groundwater is the major source of water across much of the world, particularly in rural areas in arid and semi-arid regions, but there has been very little research on the potential effec ts of climate change.

Aquifers generally are replenished by effective rainfall, rivers, and lakes. This water may reach the aquifer rapidly, through macro-pores or fissures, or more slowly by infiltrating through soils and permeable rocks overlying the aquifer. A change in the amount of effective rainfall will alter recharge, but so will a change in the duration of the recharge season. Increased wi nter rainfall—as projected under most scenarios for mid-latitudes—generally is likely to result in increased groundwater recharge. However, higher evaporation may mean that soil deficits persist for longer and commence earlier, offsetting an increase in total effective rainfall. Various types of aquifer will be recharged differently. The main types are unconfined and confined aquifers. An unconfined aquifer is recharged directly by local rainfall, rivers, and lakes, and the rate of recharge will be influenced by the permeability of overlying rocks and soils. Some examples of the effect of climate change on recharge into unconfined aquifers have been described in France, Kenya, Tanzania, Texas, New York, and Caribbean islands. Bouraoui *et al.* simulated substantial reductions in groundwater recharge near Grenoble, France, almost entirely as a result of increases in evaporation during the recharge season. Macro-pore and fissure recharge is most common in porous and ag g r egated forest soils and less common in poorly structured soils. It also occurs where the underlying geology is highly fractured or is characterized by numerous sinkholes. Such recharge can be very important in some semi-arid areas. In principle, "rapid" recharge can occur whenever it rains, so where recharge is dominated by this process it will be affected more by changes in rainfall amount than by the seasonal cycle of soil moisture variability. Sandstrom modeled recharge to an aquifer in central Tanzania and showed that a 15% reduction in r a i nfall—with no change in temperature—resulted in a 40–50% reduction in recharge; he infers that small changes in rainfall could lead to large changes in recharge and hence groundwater resources. Loaiciga *et al.* explored the effect of a range of climate change scenarios on groundwater levels in the Edwards Balcones Fault Zone aquifer in Texas, a heavily exploited aquifer largely fed by streamflow seepage. They show that, under six of the seven GCM-based scenarios used, groundwater levels and springflows would reduce s u bstantially as a result of lower streamflow. However, they use $2xCO_2$ scenarios that represent changes in temperature that are considerably greater than those projected even by the 2080s under current scenarios, so the study considerably overstates the effect of climate change in the next few decades.

Shallow unconfined aquifers along floodplains, which are most common in semi-arid and arid environments, are recharged by seasonal streamflows and can be depleted directly by evaporation. Changes in recharge therefore will be determined by changes in the duration of flow of these streams—which may locally increase or decrease—and the permeability of the overlying beds, but increased evaporative demands would tend to lead to lower groundwater storage. In semi-arid areas of Kenya, flood aquifers have been improved by construction of subsurface weirs across the river valleys, forming subsurface dams from which water is tapped by shallow wells. The thick layer of sands substantially reduces the impact of evaporation. The wells have become perennial water supply sources even during the prolonged droughts.

Sea-level rise will cause saline intrusion into coastal aquifers, with the amount of intrusion depending on local groundwater gradients. Shallow coastal aquifers are at greatest risk (on Long Island, New York, for example). Groundwater in low-lying islands therefore is very sensitive to change. In the atolls of the Pacific Ocean, water supply is sensitive to precipitation patterns and changes in storm tracks. A reduction in precipitation coupled with sea-level rise would not only cause a diminution of the harvestable volume of water; it also would reduce the size of the narrow freshwater lense. For many small island states, such as some Caribbean islands, seawater intrusion into freshwater aquifers has been observed as a result of overpumping of aquifers. Any sea-level rise would worsen the situation. It will be noted from the foregoing that unconfined aquifers are sensitive to local climate change, abstraction, and seawater intrusion. However, quantification of recharge is complicated by the characteristics of the aquifers themselves as well as overlying rocks and soils.

A confined aquifer, on the other hand, is characterized by an overlying bed that is impermeable, and local rainfall does not influence the aquifer. It is normally recharged from lakes, rivers, and rainfall that may occur at distances ranging from a few kilometers to thousands of kilometers. Recharge rates also vary from a few days to decades. The Bahariya Oasis and other groundwater aquifers in the Egyptian Desert, for example, are recharged at the Nubian Sandstone outcrops in Sudan; such aquifers may not be seriously affected by seasonal or interannual rainfall or temperature of the local area.

Attempts have been made to calculate the rate of recharge by using carbon-14 isotopes and other modeling techniques. This has been possible for aquifers that are recharged from short distances and after short durations. However, recharge that takes place from long distances and after decades or centuries has been problematic to calculate with accuracy, making es t imation of the impacts of climate change difficult. The medium through which recharge takes place often is poorly known and very heterogeneous, again challenging recharge modeling. In general, there is a need to intensify research on modeling techniques, aquifer characteristics, recharge rates, and seawater intrusion, as well as monitoring of groundwater abstractions. This research will provide a sound basis for assessment of the impacts of climate change and sea-level rise on recharge and groundwater resources.

River Flows

By far the greatest number of hydrological studies into the effects of climate change have concentrated on potential changes on streamflow and runoff. The distinction between "streamflow" and "runoff" can be vague, but in general terms streamflow is water within a river channel, usually expressed as a rate of flow past a point—typically in $m^3 s^{-1}$—whereas runoff is the amount of precipitation that does not evaporate, usually expressed as an equivalent depth of water across the area of the catchment. A simple link between the two is that runoff can be regarded as streamflow divided by catchment area, although in dry areas this does not necessarily hold because runoff generated in one part of the catchment may infiltrate before reaching a channel and becoming streamflow. Over short durations, the amount of water leaving a catchment outlet usually is expressed as streamflow; over durations of a month or more, it usually is expressed as runoff. In some c o u ntries, "runoff" implies surface runoff only (or, more pr ecisely, rapid response to an input of precipitation) and does not include the contribution of discharge from groundwater to flow, but this is a narrow definition of the term.Since the SAR, there have been many notable hydrological events—including floods and droughts—and therefore many studies into possible trends in hydrological data.

In general, the patterns found are consistent with those identified for precipitation: Runoff tends to increase where precipitation has increased and decrease where it has fallen over the past few years. Flows have increased in recent years in many parts of the United States, for example, with the greatest increases in low flows. Variations in flow from year to year have been found to be much more strongly related to precipitation changes than to temperature changes. There are some more subtle patterns, h owever. In large parts of eastern Europe, European Russia, central Canada, and California, a major—and unprecedented— shift in streamflow from spring to winter has been associated not only with a change in precipitation totals but more particularly with a rise in temperature: Precipitation has fallen as rain, rather than snow, and therefore has reached rivers more rapidly than before. In cold regions, such as northern Siberia and northern Canada, a recent increase in temperature has had little effect on flow timing because precipitation co ntinues to fall as snow.

However, it is very difficult to identify trends in hydrological data, for several reasons. Records tend to be short, and many data sets come from catchments with a long history of human intervention. Variability over time in hydrological behavior is very high, particularly in drier environments, and detection of any signal is difficult. Variability arising from low-frequency climatic rhythms is increasingly recognized, and researchers looking for trends need to correct for these patterns. Finally, land-use and other changes are continuing in many catchments, with effects that may outweigh any climatic trends. Changnon and Demissie, for example, show that human-induced changes mask the effects of climatic variability in a sample of midwest U.S. catchments. Even if a trend is identified, it may be difficult to attribute it to global warming because of other changes that are continuing in a catchment. A widespread lack of data, particularly from many developing countries, and consistent data analysis makes it impossible to obtain a representative picture of recent patterns and trends in hydrological behavior. Monitoring stations are continuing to be closed in many countries. Reconstructions of long records, stretching back centuries, are needed to understand the characteristics of natural decadal-scale variability in streamflow.

Effects of Climate Change on River Flows

By far the majority of studies into the effects of climate change on river flows have used GCMs to define changes in climate that are applied to observed climate input data to create perturbed data series. These perturbed data are then fed through a hydrological model and the resulting changes in river flows assessed. Since the SAR, there have been several global-scale assessments and a large number of catchment-scale studies. Confidence in these results is largely a function of confidence in climate change scenarios *at the catchment scale*, although Boorman and Sefton show that the use of a physically unrealistic hydrological model could lead to misleading results.

Arnell used a macro-scale hydrological model to simulate streamflow across the world at a spatial resolution of 0.5°x0.5°, under the 1961–1990 baseline climate and under several s cenarios derived from HadCM2 and HadCM3 experiments. Alcamo *et al.* also simulated the e ffects of different climate change scenarios on global river flows, showing broadly similar patterns.

All of the studies used a hydrological model to es timate the effects of climate scenarios, and all used scenarios based on GCM output. Although such studies provide extremely valuable insights into the sensitivity of hydrological systems to changes in climate, they are not assessments of the potential effects of future global warming.

A general conclusion, consistent across many studies, is that the effects of a given climate change scenario vary with catchment physical and land-cover properties and that small headwater streams may be particularly sensitive to change—as shown in northwestern Ontario.

Cold and Cool Temperate Climates

These areas are characterized by precipitation during winter falling as snow and include mountainous and low-lying regions. A major proportion of annual streamflow is formed by snow melting in spring. These areas include large parts of North America, northern and eastern Europe, most of Russia, northern China, and much of central Asia. The most important climate change effect in these regions is a change in the timing of streamflow through the year. A smaller proportion of p r ecipitation during winter falls as snow, so there is proportionately more runoff in winter and, as there is less snow to melt, less runoff during spring. Increased temperatures, in effect, reduce the size of the natural reservoir storing water during winter. In very cold climates (such as in Siberia and northern Russia), there is little change in the timing of streamflow because winter precipitation continues to fall as snow with higher temperatures.

The largest effects are in the most "marginal" snow-dominated regime areas.

The effects of climate change on the magnitude of annual runoff and flows through the year are much less consistent than the effect on streamflow timing because they depend not on the temperature increase but on the change in precipitation. In general, precipitation increases in high-latitude areas under most scenarios, but in lower latitudes precipitation may decrease. Kazcmarek *et al*., for example, show a decrease in annual runoff in Poland under a Geophysical Fluid Dynamics Laboratory (GFDL)-based scenario (by around 20% by the 2050s) but an increase under a Goddard Institute for Space Studies (GISS) scenario (by as much as 20%); in both cases, the season of maximum flow shifts from spring to winter.

Similar patterns are found for rivers in mountainous regions or draining from mountains. The Rhine and Danube, for example, would both see a reduction in spring flows and an increase in winter runoff, as would rivers draining east and west from the Rocky Mountains in North America.

Mild Temperate Climates

Hydrological regimes in these regions are dominated by the seasonal cycles of rainfall and evaporation; snowfall and snowmelt are not important. Here, climate change tends to affect the magnitude of flows in different seasons—by an amount that depends on the change in rainfall—and may lead to an exaggerated seasonal cycle, but it generally does not affect the timing of flows through the year. In the UK, for example, most scenarios result in an increase in winter runoff and, particularly in the south, a decrease in summer runoff ; similar patterns are found across most of western Europe under most scenarios. Low flows tend to occur during summer, and changes in low-flow frequency are closely related to changes in the balance between summer rainfall and summer evaporation. Across most mid-latitude temperate regions, summer rainfall would decline with global warming, leading to a reduction in low flows.

The detailed effect of a given change in climate, however, depends to a large extent on the geological characteristics of the catchment. Studies in the UK and Belgium have indicated that in catchments with considerable groundwater, changes in summer flows are largely a function of the change not in summer rainfall but in rainfall during the winter recharge season.

Arid and Semi-arid Regions

River flows in arid and semi-arid regions are very sensitive to changes in rainfall: A given percentage change in rainfall can produce a considerably larger percentage change in runoff. There have been relatively few studies in such regions since the SAR, but work has been done in southern Africa, Australia, northern China, and southern Russia.

Humid Tropical Regions

Runoff regimes in these regions are very much influenced by the timing and duration of the rainy season or seasons. Climate change therefore may affect river flows not only through a change in the magnitude of rainfall but also through possible changes in the onset or duration of rainy seasons (such as those caused by monsoon).

Lakes

Lakes are particularly vulnerable to changes in climate parameters. Variations in air temperature, precipitation, and other meteorological components directly cause changes in evaporation, water balance, lake level, ice events, hydrochemical and hydrobiological regimes, and the entire lake ecosystem. Under some climatic conditions, lakes may disappear entirely. There are many different types of lakes, classified according to lake formation and origin, the amount of water exchange, hydrochemistry, and so forth.

An important distinction is drawn between closed (endorheic) lakes, with no outflow, and exorheic lakes, which are drained by outflowing rivers. Endorheic lakes are very dependent on the balance of inflows and evaporation and are very sensitive to change in either (whether driven by climate change, climatic variability, or human interventions). This also means that they are very important indicators of climate change and can provide records of past hydroclimatic variability over a large area. Small endorheic lakes are most vulnerable to a change in climate; there are indications that even relatively small changes in inputs can produce large fluctuations in water level (and salinity) in small closed lakes in western North America.

The largest endorheic lakes in the world are the Caspian and Aral Seas, Lake Balkash, Lake Chad, Lake Titicaca, and the Great Salt Lake. Some of the largest east African lakes, including Lakes Tanganyika and Malawi, also can be regarded as practically endorheic. Changes in inflows to such lakes can have very substantial effects: The Aral Sea, for example, has been significantly reduced by increased abstractions of irrigation water upstream, the Great Salt Lake in the United States has increased in size in recent years as a result of increased precipitation in its catchment, and Qinghai Lake in China has shrunk following a fall in catchment precipitation. Many endorheic lake systems include significant internal thresholds, beyond which change may be very different. Lake Balkash, for example, currently consists of a saline part and a fresh part, connected by a narrow strait. Several rivers discharge into the fresh part, preventing salinization of the entire lake. Areduction in freshwater inflows, however, would change the lake regime and possibly lead to salinization of the freshwater part; this would effectively destroy the major source of water for a large area.

Exorheic lakes also may be sensitive to changes in the amount of inflow and the volume of evaporation. Evidence from Lake Victoria (east Africa), for example, indicates that lake levels may be increased for several years following a short-duration increase in precipitation and inflows. There also may be significant thresholds involving rapid shifts from open to closed lake conditions. Progressive southward expansion of Lake Winnipeg under postglacial isostatic tilting was suppressed by a warm dry climate in the mid-Holocene, when the north basin of the lake became closed (endorheic) and the south basin was dry. A trend of progressively moister climates within the past 5,000 years caused a return from closed to open (overflowing) lake conditions in the north basin and rapid flooding of the south basin about 1,000 years later. Other examples include Lake Manitoba, which was dry during the warm mid-Holocene. Computations of sustainable lake area under equilibrium water balance indicate that a return to dry conditions comparable to the mid-Holocene climate could cause this 24,400-km^2 lake draining a vast area from the Rocky Mountains east almost to Lake Superior to become endorheic again.

Changes in Flood Frequency

Although a change in flood risk is frequently cited as one of the potential effects of climate change, relatively few studies since the early 1990s have looked explicitly at possible changes in high flows. This largely reflects difficulties in defining credible scenarios for change in the large rainfall (or snowmelt) events that trigger flooding. Global climate models currently cannot simulate with accuracy short-duration, high-intensity, localized heavy rainfall, and a change in mean monthly rainfall may not be representative of a change in short-duration rainfall.

A few studies, however, have tried to estimate possible changes in flood frequencies, largely by assuming that changes in monthly rainfall also apply to "flood-producing" rainfall. In addition, some have looked at the possible additional effects of changes in rainfall intensity. Reynard *et al.*, for example, estimated the change in the magnitude of different return period floods in the Thames and Severn catchments, assuming first that all rainfall amounts change by the same proportion and then that only "heavy" rainfall increases. They found an increase in flood magnitudes under their wettest scenarios—even though annual runoff totals did not increase—but a decline in flood frequency under their driest scenarios.

Panagoulia and Dimou examined possible changes in flood frequency in the Acheloos basin in central Greece. Floods in this catchment derive from snowmelt, and an increase in winter precipitation—as indicated under the scenarios used—results in more frequent flood events of longer duration. The frequency and duration of small floods was most affected. Saelthun *et al.* explored the effect of fixed increases in temperature and precipitation in 25 catchments in the Nordic region. They show that higher temperatures and higher precipitation increases flood magnitudes in parts of the region where floods tended to be generated from heavy rainfall in autumn but decrease flood magnitudes where floods are generated by spring snowmelt. In some cases, the peak flood season shifts from spring to autumn. This conclusion also is likely to apply in other environments where snow and rain floods both occur.

Mirza *et al.* investigated the effects of changes in p recipitation resulting from global warming on future flooding in Bangladesh. Standardized precipitation change scenarios from four GCMs were used for the analysis. The most extreme scenario showed that for a 2°C rise in global mean temperature, the average flood discharge for the Ganges, Brahmaputra, and Meghna could be as much as 15, 6, and 19% higher, respectively.

Changes in Hydrological Drought Frequency

Droughts are considerably more difficult to define in quantitative terms than floods. Droughts may be expressed in terms of rainfall deficits, soil moisture deficits, lack of flow in a river, low groundwater levels, or low reservoir levels; different definitions are used in different sectors. A "hydrological" drought occurs when river or groundwater levels are low, and a "water resources" drought occurs when low river, groundwater, or reservoir levels impact water use. Low river flows in summer may not necessarily create a water resources drought, for example, if reservoirs are full after winter; conversely, a short-lived summer flood may not end a water resources drought caused by a prolonged lack of reservoir inflows. Water resources droughts therefore depend not only on the climatic and hydrological inputs but critically on the characteristics of the water resource system and how droughts are managed.

At the global scale, Arnell explored the change in the minimum annual total runoff with a return period of 10 years under several scenarios, based on HadCM2 and HadCM3 GCMs. He shows that the pattern of this measure of "low flow" (which is relatively crude) changes in a similar way to average annual runoff but that the percentage changes tend to be larger. Arnell mapped a different index of low flow across Europe—the average summed difference between streamflow and the flow exceeded 95% of the time, while flows are below this threshold—under four scenarios. The results suggest a reduction in the magnitude of low flows under most scenarios across much of western Europe, as a result of lower flows during summer, but an a m elioration of low flows in the east because of increased winter flows. In these regions, however, the season of lowest flows tends to shift from the current winter low-flow season toward summer.

Döll *et al.* also modeled global runoff at a spatial resolution of 0.5°x0.5°, not only for average climatic conditions but also for typical dry years. The annual runoff exceeded in 9 years out of 10 (the 10-year return period "drought" runoff) was derived for each of more than 1,000 river basins covering the whole globe. Then the impact of climate change on these runoff values was computed by scaling observed temperature and precipitation in the 1-in-10 dry years with climate scenarios of two different GCMs, ECHAM4/OPYC3 and GFDL-R15. Climate variability was assumed to remain constant. For the same GHG emission scenario, IS92a, the two GCMs compute quite different temperature and more so precipitation changes. With the GFDL scenario, runoff in 2025 and 2075 is simulated to be higher in most river basins than with the ECHAM scenario. The 1-in-10 dry year runoff is computed to decrease between the present time (1961–1990 climate) and 2075 by more than 10% on 19% (ECHAM) or 13% (GFDL) of the global land area and to increase by more than 50% on 22% (ECHAM) or 49% (GFDL) of the global land area. These results underline the high sensitivity of computed future runoff changes to GCM calculations.

There have been several other studies into changes in low flow indicators at the catchment scale. Gellens and Roulin, for example, simulated changes in low flows in several Belgian catchments under a range of GCM-based scenarios. They show how the same scenario could produce rather different changes in different catchments, depending largely on the catchment geological conditions. Catchments with large amounts of groundwater storage tend to have higher summer flows under the climate change scenarios considered because additional winter rainfall tends to lead to greater groundwater recharge (the extra rainfall offsets the shorter recharge season). Low flows in catchments with little storage tend to be reduced because these catchments do not feel the benefits of increased winter recharge. Arnell and Reynard found similar results in the UK. The effect of climate change on low flow magnitudes and frequency therefore can be considered to be very significantly affected by catchment geology (and, indeed, storage capacity in general). Dvorak *et al.* also showed how changes in low flow measures tend to be proportionately greater than changes in annual, seasonal, or monthly flows.

Water Quality

Water in rivers, aquifers, and lakes naturally contains many dissolved materials, depending on atmospheric inputs, geological conditions, and climate. These materials define the water's chemical characteristics. Its biological characteristics are defined by the flora and fauna within the water body, and temperature, sediment load, and color are important physical characteristics. Water "quality" is a function of chemical, physical, and biological characteristics but is a value-laden term because it implies quality in relation to some standard. Different uses of water have different standards. Pollution can be broadly defined as deterioration of some aspect of the chemical, physical, or biological characteristics of water (its "quality") to such an extent that it impacts some use of that water or ecosystems within the water. Major water pollutants include organic material, which causes oxygen deficiency in water bodies; nutrients, which cause excessive growth of algae in lakes and coastal areas—known as eutrophication (leading to algal blooms, which may be toxic and consume large amounts of oxygen when decaying); and toxic heavy metals and organic compounds. The severity of water pollution is governed by the intensity of pollutants and the assimilation capacity of receiving water bodies—which depends on the physical, chemical, and biological characteristics of streamflow— but not all pollutants can be degraded, however.

Chemical river water quality is a function of the chemical load applied to the river, water temperature, and the volume of flow. The load is determined by catchment geological and land-use characteristics, as well as by human activities in the catchment: Agriculture, industry, and public water use also may result in the input of "polluting" substances. Agricultural inputs are most likely to be affected by climate change because a changing climate might alter agricultural practices. A changing climate also may alter chemical processes in the soil, including chemical weathering. Avila *et al.* simulated a substantial increase in base cation weathering rates in Spain when temperature and precipitation increased (although if precipitation were reduced, the effects of the higher temperature were offset). This, in turn, resulted in an increase in concentrations of base cations such as calcium, sodium, and potassium and an increase in streamwater alkalinity. Warmer, drier conditions, for example, promote mineralization of organic nitrogen and thus increase the potential supply to the river or groundwater. Load also is influenced by the processes by which water reaches the river channel. Nitrates, for example, frequently are flushed into rivers in intense storms following prolonged dry periods.

River water temperature depends not only on atmospheric t emperature but also on wind and solar radiation. River water temperature will increase by a slightly les ser amount than air temperature, with the smallest increases in catchments with large contributions from groundwater. Biological and chemical processes in river water are dependent on water temperature: Higher temperatures alone would lead to increases in concentrations of some chemical species but decreases in others. Dissolved oxygen concentrations are lower in warmer water, and higher temperatures also would encourage the growth of algal blooms, which consume oxygen on decomposition.

Streamwater quality, however, also will be affected by streamflow volumes, affecting both concentrations and total loads. Carmichael *et al.*, for example, show how higher temperatures and lower summer flows could combine in the Nitra River, Slovakia, to produce substantial reductions in d issolved oxygen concentrations. Research in Finland indicates that changes in stream water quality, in terms of eutrophication and nutrient transport, are very dependent on changes in streamflow. For a given level of inputs, a reduction in streamflow might lead to increases in peak concentrations of certain chemical compounds. Cruise *et al.* simulated increased concentrations of nitrate in the southeast United States, for example, but the total amount transported from a catchment might decrease. Hanratty and Stefan simulated reductions in nitrate and phosphate loads in a small

Minnesota catchment, largely as a result of reductions in runoff. Alexander *et al.* suggest that nutrient loadings to receiving coastal zones would vary primarily with streamflow volume. Increased streamflow draining toward the Atlantic coast of the United States under many scenarios, for example, would lead to increased nutrient loadings. An increased frequency of heavy rainfall would adversely affect water quality by increasing pollutant loads flushed into rivers and possibly by causing overflows of sewers and waste storage facilities. Polluting material also may be washed into rivers and lakes following inundation of waste sites and other facilities located on floodplains.

Water temperature in lakes responds to climate change in more complicated ways because thermal stratification is formed in summer, as well as in colder regions in winter. Meyer *et al.* evaluated the effect of climate change on thermal stratification by simulation for hypothetical lakes. They show that lakes in subtropic zones (about latitude 30 to 45°) and in subpolar zones (latitude 65 to 80°) are subject to greater relative changes in thermal stratification patterns than mid-latitude or equatorial lakes and that deep lakes are more sensitive than shallow lakes in the subtropic zones. Hostetler and Small simulated potential impacts on hypothetical shallow and deep lakes across North America, showing widespread increases in lake water temperature slightly below the increase in air temperature in the scenarios used. The greatest increases were in lakes that were simulated to experience substantial reductions in the duration of ice cover; the boundary of ice-free conditions shifted northward by 10° of latitude or more (1,000 km). Fang and Stefan show by simulation that winter stratification in cold regions would be weakened and the anoxic zone would disappear. Observations during droughts in the boreal region of northwestern Ontario show that lower inflows and higher temperatures produce a deepening of the thermocline.

The consequences of these direct changes to water quality of polluted water bodies may be profound, as summarized by Varis and Somlyody for lakes. Increases in temperature would deteriorate water quality in most polluted water bodies by increasing oxygen-consuming biological activities and decreasing the saturation concentration of dissolved oxygen. Hassan *et al.* employed a downscaled climate model combined with GCM output to predict future stratification for Suwa Lake, Japan, on a daily basis, as well as for the prolonged summer stratification period. They predict increased growth of phytoplankton and reduced dissolved oxygen concentrations at different depths in the lake. Analysis of past observations in Lake Biwa in Japan suggests that dissolved oxygen concentrations also tend to reduce when air (and lake water) temperature is higher.

Water quality in many rivers, lakes, and aquifers, however, is heavily dependent on direct and indirect human activities. Land-use and agricultural practices have a very significant effect on water quality, as do management actions to control point and nonpoint source pollution and treat wastewaters d i scharged into the environment. In such water bodies, future water quality will be very dependent on future human activities, including water management policies, and the direct effect of climate change may be very small in relative terms.

Confidence in estimates of change in water quality is determined partly by climate change scenarios (and their effects on streamflow), but additional uncertainty is added by current lack of detailed understanding of some of the process interactions involved.

Glaciers and Small Ice Caps

Valley glaciers and small ice caps represent storages of water over long time scales. Many rivers are supported by glacier melt, which maintains flows through the summer season. The state of a glacier is characterized by the relationship between the rate of accumulation of ice (from winter snowfall) and the rate of ablation or melt. Most, but not all, valley glaciers and small ice caps have been in general retreat since the end of the Little Ice Age, between 100 and 300 years ago—for example, in Switzerland, Alaska, the Canadian Rockies, east Africa, South America, the arid region of northwest China, and tropical areas as a whole. Temperature appears to be the primary control, and rates of retreat generally are accelerating. The World Glacier Monitoring Service.

The effect of future climate change on valley glaciers and small ice caps depends on the extent to which higher temperatures are offset by increased winter accumulation. At the global scale, Gregory and Oerlemans simulate a general decline in valley glacier mass (and consequent rise in sea level), i n d icating that the effects of higher temperatures generally are more significant than those of additional winter accumulation. Model studies of individual glaciers have shown general retreat with global warming. Wallinga and van de Wal and Haerberli and Beniston, for example, both simulated retreat in Alpine glaciers with higher temperatures and changes in winter accumulation. Davidovich and Ananicheva's simulation results show retreat of Alaskan glaciers but also a substantial increase in mass exchange (and therefore rate of movement) as a result of increased winter accumulation.

Oerlemans *et al.* simulated the mass balance of 12 valley glaciers and small ice sheets distributed across the world. They found that most scenarios result in retreat (again showing that temperature changes are more important than precipitation changes) but showed that it was very difficult to generalize results because the rate of change depends very much on glacier hypsometry (i.e., variation in altitude across the glacier). Their simulations also show that, in the absence of a change in p recipitation, a rise in temperature of 0.4°C per decade would virtually eliminate all of their study glaciers by 2100, but a rise of 0.1°C per decade would "only" lead to a reduction in glacier volume of 10–20%.

Tropical glaciers are particularly exposed to global warming. Kaser *et al.* show that the equilibrium line altitude (ELA)—the line separating the accumulation zone from the ablation zone—of a tropical glacier is relatively more sensitive to changes in air temperature than that of a mid-latitude glacier. This is because of the lack of seasonality in tropical temperatures and the fact that ablation is significant year-round. To illustrate, a 1°C rise in temperature during half of the year only will have a direct impact on total ablation, annual mass balance, and ELA of a tropical glacier. In the case of a mid-latitude glacier, this increase may occur during winter when temperatures may be well below freezing over much (if not all) of the glacier. As a result, there may be no significant change in ablation or position of the ELA, even though the annual temperature will have increased.

Glacier retreat has implications for downstream river flows. In rivers fed by glaciers, summer flows are supported by glacier melt (with the glacier contribution depending on the size of the glacier relative to basin area, as well as the rate of annual melt). If the glacier is in equilibrium, the amount of precipitation stored in winter is matched by melt during summer. However, as the glacier melts as a result of global warming, flows would be expected to increase during summer—as water is released from long-term storage—which may compensate for a reduction in precipitation. As the glacier gets smaller and the volume of melt reduces, summer flows will no longer be supported and will decline to below present levels. The duration of the period of increased flows will depend on glacier size and the rate at which the glacier melts; the smaller the glacier, the shorter lived the increase in flows and the sooner the onset of the reduction in summer flows.

River Channel Form and Stability

Patterns of river channel erosion and sedimentation are determined largely by variations in streamflow over time—in particular, the frequency of floods. There is considerable literature on past changes in streamflow—caused by human influences or natural climatic variability—and associated river channel changes but very little on possible future channel changes. This largely reflects a lack of numerical models to simulate erosion and sedimentation processes; assessments of possible future channel changes that have been made have been inferred from past changes. In northern England, for example, Rumsby and Mackin show that periods with large numbers of large floods are characterized by channel incision, whereas periods with few floods were c haracterized by lateral reworking and sediment transfer. Increased flooding in the future therefore could be associated with increased channel erosion.

The density of the drainage network reflects the signature of climate on topography. Moglen *et al.* show that drainage density is sensitive to climate change but also that the direction of change in density depends not only on climate change but also on the current climate regime.

Hanratty and Stefan simulated streamflow and sediment yield in a small catchment in Minnesota. The scenario they used produced a reduction in sediment yield, largely as a result of reduced soil erosion, but their confidence in the model results was low. In fact, the lack of physically based models of river channel form and sediment transport means that the c onfidence in estimates of the effect of climate change on river channels is low in general.

Climate Change and Climatic Variability

Even in the absence of a human-induced climate change, hydrological behavior will vary not only from year to year but also from decade to decade. Hulme *et al.* simulated streamflow across Europe under four climate change scenarios for the 2 050s (based on four differents imulations from the HadCM2 climate model) and seven scenarios representing different 30-year climates extracted from a long run of the HadCM2 model with no GHG forcing. They show that natural multi-decadal (30-year) variability in average annual runoff is high across most of Europe and that this natural variability in runoff in mid-latitude Europe is greater than the simulated signal of climate change. In northern and southern Europe, the magnitude of climate change by the 2050s is greater than the magnitude of natural variability. However, the spatial patterns of climate change and climatic variability are very different, with a much more coherent (usually north-south) pattern in the climate change signal. Nevertheless, the results indicate that, for individual catchments in certain areas, the magnitude of climate change effects on some indicators of streamflow may be smaller than natural climatic variability for several decades, whereas in other areas, the climate change signal will be larger than past experience.

Effects on Water Withdrawals

The consequences of climate change for water resources depend not only on possible changes in the resource base—as indicated —but also on changes in the *demand*, both human and environmental, for that resource.

It must be noted that "demand" in its economic sense means willingness to pay for a particular service or commodity and is a function of many variables—particularly price, income (for households), output (for industries or agriculture), family composition, education levels, and so forth. The usefulness of the demand function is found in the ability to predict the effects of changes in causal variables and in measurement of the demanding party's "willingness to pay" as a measure of gross benefits to the demanding party of various quantities. This willingness to pay is measured as the area under the demand function in the price-quantity plane. The quantities *actually* purchased (the quantities of water withdrawn or used) over time are the result of the interaction of factors affecting demand as defined above and conditions of supply (or availability). Thus, for example, the fact that the quantity purchased over time increases could be the result of falling costs of supply (a shift in the supply curve) rather than an increase in demand (shift in the demand curve). In this section, the term "demand" often is used as a synonym for "requirements;" this reflects usage of the term in large parts of the water sector.

Demands can be classified along two dimensions: instream or offstream, and consumptive or nonconsumptive. Instream demands use water within the river channel (or lake) and do not involve withdrawal. Examples include ecosystem uses, navigation, hydropower generation, recreation, and use of the water course for waste assimilation. Offstream demands extract water from the river channel, lake, or aquifer. They include domestic, industrial, and agricultural demands, as well as extractions for industrial and power station cooling. These demands can be consumptive or nonconsumptive. Consumptive demands "use" the water so it cannot be entirely returned to the river; nonconsumptive demands return the water to the river, although it may be returned to a different catchment or at a different quality. The primary consumptive demands are for irrigation and some types of industrial cooling (where the water is evaporated to the atmosphere rather than returned to the river).

World Water Use

Agricultural use—primarily for irrigation—is by far the largest proportion, accounting in 1995 for 67% of all withdrawals and 79% of all water consumed. Municipal, or domestic, use represents only about 9% of withdrawals. There are large differences, of course, between continents, with the greatest absolute volume of irrgation withdrawals in Asia.

Over the past few years there have been many projections of future water withdrawals; virtually all have overestimated the actual rate of increase. The central projection represents a "Conventional Development Scenario" (CDS), with "best-guess" estimates of future population growth, economic development, and water-use intensity. The upper and lower lines represent high and low cases, where the assumed rates of growth are altered. Under the core CDS, global water withdrawals would increase by about 35% over 1995 values by 2025, with low and high estimates of 23 and 49%, respectively. The greatest rates of growth are projected to be in Africa and the Middle East, with the lowest growth in developed economies. These projections are very dependent not only on the assumed rate of population growth but also on the different assumed rates of water usage. They do not take potential climate change into account.

The amount of municipal water withdrawals is dependent on the number of urban dwellers, the level of development (related to the availability of a piped distribution network and sewage systems, amongst other things), income levels, and price where actually applied to water. Per capita withdrawals in many developed countries are falling, and this may more than offset an increasing population: Total municipal withdrawals in the United States, for example, are falling largely as a result of increasing prices, conservation education provided by water utilities, and increasing use of water-efficient appliances. However, municipal withdrawals in developing countries can be expected to increase with living standards; under the CDS, per capita withdrawals in these countries are projected to rise toward those in more economically developed countries. Rapid urbanization in developing countries is expected to lead to very substantial increases in total municipal water withdrawals.

The industrial sector currently accounts for approximately 20% of current withdrawals. This water is used primarily either in processing (food processing or heavy industry) or for cooling; the relative proportions vary between countries. Future industrial water use is expected to rise substantially as industrial development continues, but the increase probably will be less than the increase in industrial production as water is used with greater efficiency (using less per unit of production, for example, or relocating power generation plants to coastal areas to use seawater for cooling). Under the CDS, total industrial water use increases; this increase is concentrated largely in Asia and Latin America.

The amount of water used for agriculture is dependent primarily on the level of irrigation development, the pricing of water, and the reliability of supply. Future irrigation uses are a function of the rate of expansion of irrigated land, irrigation efficiency, and pricing practices. Efficiency is projected to improve—partly as a result of changes in the cost of water and partly as a result of technological developments—but this may be more than offset by increases in the area under irrigation. The amount of increase, however, is very uncertain; it depends on, among other things, assumed rates of population growth (increasing demand) and assumed changes in world agricultural markets. Previous estimates of future water use have tended to overestimate the rate of increase because the rate of expansion of irrigated land has been overestimated. The expanded use of pricing to reflect water scarcity is being strongly promoted by international organizations and will affect the quantities used. However, many regions that historically have been dependent on rainfall are using *supplemental irrigation* —a factor that will increase irrigation use.

Estimates of future water withdrawals are notoriously uncertain, largely reflecting uncertainties in the future rate of population and economic growth. There is an analogy here with GHG emissions scenarios. Also important, however, are possible changes in the way water is priced. Much agricultural water, for example, currently is heavily subsidized, and a shift toward a more "economic" price for water is likely to have a very significant effect on use. The World Bank and other economic development agencies are pushing for major reforms in the way water is priced and sold, as well as the use of water markets as an efficient way of reallocating existing supplies and motivating efficiency.

Sensitivity of Demand to Climate Change

Climate change is another potential influence on the demand for water. Municipal demand is related to climate to a certain extent. Shiklomanov notes different rates of use in diffe r ent cl imate zones, although in making comparisons between cities it is difficult to account for variation in nonclimatic controls. The sensitivity of municipal demand to climate change is likely to be very dependent on the uses to which the water is put. The most sensitive areas are increased personal washing and— more importantly in some cultures—increased use of water in the garden and particularly on the lawn. Studies in the UK suggest that a rise in temperature of about 1.1°C by 2025 would lead to an increase in average per capita domestic demand of approximately 5%—in addition to nonclimatic trends—but would result in a larger percentage increase in peak demands (demands for garden watering may be highly concentrated). Boland estimated the effects of climate change on municipal demand in Washington, D.C., under a range of different water conservation policies.

Industrial use for processing purposes is insensitive to climate change; it is conditioned by technologies and modes of use. Demands for cooling water, however, may be affected by climate change. Increased water temperatures will reduce the efficiency of cooling, perhaps necessitating increased abstraction (or, of course, changes in cooling technologies to make them more efficient).

Agricultural demand, particularly for irrigation water, is co nsiderably more sensitive to climate change. There are two potential effects. First, a change in field-level climate may alter the need for and timing of irrigation: Increased dryness may lead to increased demands, but demands could be reduced if soil moisture content rises at critical times of the year. Döll and Siebert applied a global irrigation water-use model with a spatial resolution of 0.5°x0.5° to assess the impact of climate change on net irrigation requirements per unit irrigated area, with a climate change scenario based on the ECHAM4 GCM. Under this scenario— and similarly under the corresponding HadCM3 scenario—net irrigation requirements per unit irrigated area generally would decrease across much of the Middle East and northern Africa as a result of increased precipitation, whereas most irrigated areas in India would require more water. The extra irrigation requirements per unit area in most parts of China would be small; the HadCM3 scenario leads to a greater increase in northern China. Other climate models would give different indications of regional changes in irrigation requirements. On the global scale, increases and decreases in net irrigation requirements largely cancel, and there is less difference between different climate models; under two scenarios considered by Döll and Siebert, global net irrigation requirements would increase, relative to the situation without climate change, by 3.5–5% by 2025 and 6–8% by 2075. Actual changes in withdrawals would be dependent on changes in the efficiency of irrigation water use.

The second potential effect of climate change on irrigation demand is through increasing atmospheric CO_2 concentrations. Higher CO_2 concentrations lower plant stomatal conductance, hence increase WUE; but as indicated, this may be offset to a large extent by increased plant growth.

Hatch *et al.* assessed irrigation water requirements in Georgia, USA, using a climate change scenario derived from HadCM2. This scenario produced increased rainfall in most seasons, which, together with a shorter growing season and the assumed effect of CO_2 enrichment, resulted in a *decrease* in irrigation demand, ranging from just 1% by 2030 for soybean to as much as 20% by 2030 for corn. Along the Gulf Coast of the United States, however, the same scenario implies an increase in irrigation demands. Strzepek *et al.* also simulated decreases in irrigation requirements across the U.S. cornbelt under two of three scenarios (with the decrease depending on assumed irrigation use efficiency) but an increase under the third scenario. These three studies together indicate considerable uncertainty in estimated future irrigation withdrawals.

Impacts on Water Resources and Hazards

The preceding sections have assessed the potential effect of c limate change on river flows, groundwater recharge and other biophysical components of the water resource base, and demands for that resource. The consequences, or impacts, of such changes on risk or resource reliability depend not only on the biophysical changes in streamflow, recharge, sea-level rise, and water quality but also on the characteristics of the water management system. In practice, of course, the *actual* impacts of climate change will be rather different because water managers will make incremental or autonomous adaptations to change—albeit on the basis of imperfect know ledge—and the impact of change will be a function of adaptation costs and residual impacts. However, very few s t udies have incorporated deliberate adaptation strategies, and studies that do not consider adaptation provide a base case for assessing the magnitude of the climate change "problem." More significant, some studies

have not accounted for nonclimatic changes in the way water resources are managed or systems are operated and have applied the future climate to the present management system. This is unrealistic, but the extent of adaptation by many water managers is uncertain. It is important to assess the effect of climate change by, say, the 2050s in the context of the water management system that would exist by then in the absence of climate change—considering, for example, changes in demand or legislative requirements.

The sensitivity of a water resource system to climate change is a function of several physical features and, importantly, societal characteristics. Physical features that are associated with m a ximum sensitivity include:

- A current hydrological and climatic regime that is m arginal for agriculture and livestock
- Highly seasonal hydrology as a result of either seasonal precipitation or dependence on snowmelt
- High rates of sedimentation of reservoir storage
- Topography and land-use patterns that promote soil erosion and flash flooding conditions
- Lack of variety in climatic conditions across the territory of the national state, leading to inability to relocate activities in response to climate change.

Societal characteristics that maximize susceptibility to climate change include:

- Poverty and low income levels, which prevent long-term planning and provisioning at the household level
- Lack of water control infrastructures

- Lack of maintenance and deterioration of existing infrastructure
- Lack of human capital skills for system planning and management
- Lack of appropriate, empowered institutions
- Absence of appropriate land-use planning
- High population densities and other factors that inhibit population mobility
- Increasing demand for water because of rapid population growth
- Conservative attitudes toward risk [unwillingness to live with some risks as a tradeoff against more goods and services (risk aversion)]
- Lack of formal links among the various parties involved in water management.

A Global Perspective

There are several indicators of water resource stress, including the amount of water available per person and the ratio of volume of water withdrawn to volume of water potentially available. When withdrawals are greater than 20% of total renewable resources, water stress often is a limiting factor on development; withdrawals of 40% or more represents high stress. Similarly, water stress may be a problem if a country or region has less than 1,700 m^3 yr^{-1} of water per capita. Simple numerical indices, however, give only partial indications of water resources pressures in a country or region because the consequences of "water stress" depend on how the water is managed.

At the global scale, assessments of water stress usually are made by country because that is the unit at which water-use data generally are available. In 1990, approximately one-third of the world's population lived in countries using more than 20% of their water resources, and by 2025 about 60% of a larger total would be living in such stressed countries, in the absence of climate change, largely because population growth. Arnell estimates the effect of a number of climate change scenarios on national water resource availability and compares this with estimated future demands for water (increasing following the CDS outlined. There is considerable variability between scenarios, essentially reflecting how resources change in populous countries, but by the 2020s the table indicates that about 0.5 billion people could see increased water resources stress as a result of climate change. Significant geographic variations are hidden. Under most of the scenarios considered, climate change increases stresses in many countries in southern and western Africa and the Middle East, whereas it ameliorates stresses in parts of Asia. Alcamo *et al.* found broadly similar results.

Climate change tends to have a small effect relative to population growth, and the range of magnitudes of effect between scenarios also is little changed; the effects are still uncertain. For most of the example countries, climate change may result in either an increase or a decrease, although for some the climate change signal is more consistent (reductions in South Africa, Cyprus, and Turkey, for example, and increases in China). Note that these figures represent national averages, and different parts of each country may be differently affected.

Catchment and System Case Studies

Although there have been many assessments of the effect of climate change on river flows and (to a much lesser extent) groundwater recharge, there have been few published quantitative studies into impacts on real water resource systems.

Perhaps the most comprehensive set of studies conducted to date was published by Lettenmaier *et al.*. This study looked at potential climate change impacts on six dimensions of water resource (power generation, municipal water supply, recreation, flood defense, navigation, and environmental flow regulation) in six major U.S. basins, using realistic models of the water system operation and scenarios for possible future nonclimatic changes in demands and objectives (not all the six dimensions were assessed in each basin). The results clearly show considerable variability between scenarios and catchments; they also show that climate change might produce water resources benefits. The results also indicate that, in most sectors and basins, the impacts of different demand and operational assumptions by 2050 are greater than, or of similar magnitude to, the potential impacts of climate change.

Kaczmarek *et al.* assessed the impact of climate change on the water supply system in the Warta River basin in Poland, looking at two climate change scenarios in the context of increasing demand for water (particularly from irrigation). In the absence of climate change, they show that there would be supply problems in part of the system by 2050, simply because of the increase in demand. Under one of the scenarios, inflows to supply reservoirs would increase sufficiently to prevent s upply problems; under the other scenario, the risk of shortage would increase substantially (the probability of an annual deficit of 10% would increase from 4 to ~25%, for example). Kaczmarek *et al.* also looked at the feasibility of one adaptation option—transferring water from one reservoir to another—and showed how it could lessen the likelihood of shortage.

The River Rhine is a very important transport route within Europe. Grabs considered the effect of two climate change scenarios on navigation opportunities, having translated climate into streamflow by using a catchment water balance model.

Very few explore the costs of these impacts, primarily because of difficulties in deciding the basis for c a lculation. Are the costs of climate change equal to the cost of continuing to provide the current standard of service? Are the costs of services foregone (in terms of extra flood damages or reduced use of water), or are they incurred in providing services at a new economically-optimum level? In other words, estimates of the cost of climate change must consider explicitly the m e asures used to adapt to that change, and the economic costs of climate change will depend on the adaptation strategies adopted. Carmichael *et al.* present one of the few studies that has tried to cost the implications of climate change. They investigated the treatment costs necessary to maintain a given water quality standard (expressed in terms of dissolved oxygen content) in a river in Slovakia and calculated the least costly treatment under the present hydrological regime and under one scenario for the 2020s. They showed that costs would be little different if the aim were to meet a 4 ppm dissolved oxygen target under average summer conditions but would rise by a factor of about 14 (at current prices) if the aim were to meet the same target under low-flow conditions, even taking a least-cost approach.

Aggregated estimates of the cost of impacts of climate change on water resources have been prepared for Spain, the UK, and the United States. Ayala-Carcedo and Iglesias-Lopez estimate that the reduction in water supplies under one scenario would cost nearly US$17 billion between 2000 and 2060, or about US$280 million yr^{-1} in terms of increased expenditure to maintain supplies and lost agricultural production. A study in the UK estimated the costs of climate change for water supply and flood protection. The study assumes that the same change in water availability occurred across all of Britain—which probably overstates the costs because many parts of Britain are projected to have increased runoff—and estimated costs on the basis of standardized costs per unit of water. The study does not consider the feasibility of each of the potential adaptations. The cost of demand management measures increases substantially for large reductions in demand because more expensive technologies are needed. Note that a 5% reduction in demand represents just more than half the water of a 5% increase in supply; reducing domestic demand by 20% has a similar effect to increasing supply by 10%. The ERM study assumes that annual riverine flood damages would increase, because of increased flooding, by about US$80–170 million yr^{-1} over the next 30 years (compared to a current figure of about US$450 million), and the average annual cost of building structural works to prevent this extra flooding would be about US$40 million.

There have been two sets of estimates of the aggregate cost of climate change for water resources in the United States, using different approaches. Hurd *et al.* examined four river basins under nine climate change scenarios (defining fixed changes in temperature and precipitation) and extrapolated to the United States as a whole. Their study uses detailed economic and hydrological modeling and suggests that the largest costs would arise through maintaining water quality at 1995 standards—US$5.68 billion yr^{-1} (1994 US$) by 2060 with a temperature increase of 2.5°C and a 7% increase in precipitation— and through lost hydropower production (US$2.75 billion yr^{-1} by 2060, under the same scenario). Costs of maintaining public water supplies would be small, and although loss of irrigation water would impact agricultural users, changed cropping and irrigation patterns would mean that the economic losses to agriculture would be less than US$0.94 billion yr^{-1} by 2060. However, this study extrapolates from the four study catchments to the entire United States by assuming that the same climate change would apply across the whole country.

Frederick and Schwarz take a different approach, looking at 18 major water resource regions and 99 assessment subregions, with two climate change scenarios for the 2030s based on climate model simulations. Water scarcity indices were developed for each assessment subregion, comparing scarcities under "desired streamflow conditions" and "critical streamflow conditions" on the demand side with "mean streamflows" and "dry-condition streamflows" on the supply side. These indices played a key role in determining the costs of meeting various streamflow targets. A supply - demand balance in each region is achieved through supply- and demand-side measures, each of which has an assumed unit cost. Three strategies were defined for each region: "environmental," focusing on protecting the environment; "efficient," maintaini n g supplies to users; and "institutional," placing limits on changes in environmental indicators and the area of irrigation. The total national cost of climate change was determined under each strategy by aggregating least-cost measures in each subregion. The costs are considerably greater under the drier CGCM1 scenario than under the wetter HadCM2 scenario (which, in fact, implies a benefit), and they vary with management strategy. Costs under the drier scenario are considerably higher than those estimated by Hurd *et al.*, reflecting partly the different approaches used and partly the spatial variability in the effect of climate change considered by Frederick and Schwarz.

It is very difficult to draw quantitative conclusions about the impacts of climate change, for several reasons. Different studieshave used different methodologies and different scenarios, but, most important, different systems respond very differently to climate change. It is possible, however, to make some qualitative generalizations:

- In systems with large reservoir capacity, changes in resource reliability may be proportionately smaller than changes in river flows.
- The potential impacts of climate change must be c onsidered in the context of other changes that affect water management. Few studies have explicitly compared climate change with other pressures, but in many environments it is likely that over a time horizon of less than 20 years, climate change impacts will be very small relative to other pressures. This will depend on the system.
- The implications of climate change are likely to be greatest in systems that currently are highly stressed.

By far the majority of studies of the impact of climate change on water resources have concentrated on human aspects of the water environment. Only a very few have considered impacts on the aquatic environment. Some of these studies are considered, but it must be remembered that water resources systems in many parts of the world increasingly are being managed to maintain instream and wetland ecosystems. This either increases effective water demand or decreases water availability.

Confidence in estimated quantitative impacts of climate change on water resources generally is low, reflecting initial confidence in climate change scenarios and low confidence in estimates of future pressures on water resources (as a result of factors such as changes in demand or legislative requirements).

Adaptation Options and Management Implications

The preceding sections have assessed the possible effects of climate change on the water resource base and on the demand for water, as well as the potential impacts on water users. Most published studies have looked at impacts in the absence of planned adaptation to climate change, and the few studies that have tried to cost impacts have had to make assumptions about adaptation.

Water management has always adapted to change (especially following extreme events or in response to increased demand), and climate change is just one of the pressures facing water managers. Other pressures include increasing demands for water resources or protection against hazard, changing water management objectives (which recently have included increasing recognition of the importance of meeting environmental needs as well as those of offstream demands), changing water management technologies, and altered legislative environments.

Table 1. National average annual cost of maintaining water supply-demand balance in the USA. Values in 1994 US$ billion.

Management Strategy	*HadCM2*	*CGCM1*
"Efficient"	-4.7	105
"Environmental"	-4.7	251
"Institutional"	not calculated	171

It is important to distinguish between development of adaptive options for meeting changing demands and resources and assessment of the abilities of a given water management agency (interpreted broadly) actually to adapt to climate change. Over the years, a wide range of adaptive techniques has been developed, largely in response to the need to meet increased demands. Broad distinctions can be drawn among "supply-side" adaptive techniques (changing structures, operating rules, and institutional arrangements) and "demand-side" t echniques (which change the demand for water or protection against risk and include institutional changes as well). Examples of supply-side adaptations include increasing flood defenses, building weirs and locks to manage water levels for navigation, and modifying or extending infrastructure to collect and distribute water to consumers. Demand-side techniques include water demand management (such as encouraging water-efficient irrigation and water pricing initiatives), changing water allocations, and nonstructural flood management measures (such as land-use controls). Distinctions also can be drawn between anticipatory and reactive actions. The former are taken in advance of some change, the latter in response to a change. Reactive actions include short-term operational adaptations, such as temporary exploitation of new sources, and longer term measures. Amajor flood or drought, for example, often triggers a change in water management. However, although many adaptive options do exist, knowledge of these options and the expertise of officials to execute them may be limited in some situations.

The optimum extent of adaptation can be characterized in terms of the benefits and costs of adaptation. The extremes of adaptation are "no adaptation" and "adaptation sufficient to eliminate all effects" (which usually is not physically possible). The optimum level of adaptation minimizes the combined costs of adaptation and residual negative effects, with the most cost-effective steps taken first.

Water managers long have had access to many techniques for assessing options and implementing adaptive strategies. However, the techniques used have changed over time and vary between countries, and they are very much influenced by institutional arrangements in place in a country. Factors that affect adaptive capacity in a country include institutional capacity, wealth, management philosophy (particularly m a nagement attitudes toward supply-side versus demand-side strategies, as well as "sustainable" management), planning time scale, and organizational arrangements (adaptation will be harder, for example, when there are many different "managers" involved or where water managers do not have sound professional guidance).

Water Management Options

Most of strategies are being adopted or considered in many countries in the face of increasing demands for water resources or protection against risk. In the UK, for example, water supply companies currently are pursuing the "twin track" of demand management and supply management in response to potential increases in demand for water (although there is a conflict between different parts of the water management system over the relative speeds with which the two tracks should be followed). These management strategies also are potentially feasible in the face of climate change. Nowhere, however, are water management actions being taken explicitly and solely to cope with climate change, although in an increasing number of countries climate change is being considered in assessing future resource management. In the UK, for example, climate change is one of the factors that must be considered by water supply companies in assessing their future resource requirements—although companies are highly unlikely to have new resources justified at present on climate change alone.

The continuing debate in water management is between the practicalities and costs of supply-side ve rsus demand-side options, and this debate is being pursued indepedently of climate change. The tide is moving toward the use of demand-side options because they are regarded as being more environmentally sustainable, cost-effective, and flexible. "Smart" combinations of supply-side and demand-side approaches are needed, although in many cases new supply-side infrastructure may be necessary. This is particularly the case in developing countries, where the challenge often is not to curb demand but to meet minimum human health-driven standards.

There do appear, however, to be numerous "no regret" policies that warrant immediate attention. In this context, a "no regret" policy is one that would generate net social benefits regardless of whether there was climate change. The effect of successful demand-side policies is to reduce the need for supply augmentation, although they may not prevent such needs entirely if changes are large. Such policy changes represent the minimum package of "anticipatory policy changes" in response to climate change.

Implications of Climate Change for Water Management Policy

Climate change exaggerates current pressures in water m a nagement—adding to the debate on sound management strategies—and adds a new component. This new component relates to uncertainty in climate change: How can water m anagement efficiently adapt to climate change, given that the magnitude (or possibly even the direction) of change is not known? Conventionally, water resource managers assume that the future resource base will be the same as that of the past and therefore that estimates of indices such as average reservoir yield or probable maximum flood that are based on past data will apply in the future. There are two issues: assessing alternatives in the face of uncertainty and making decisions on the basis of this assessment.

Techniques for assessing alternatives include scenario analysis and risk analysis. Scenario analysis is central to climate change impact assessment, but it is not widely used in water resource assessment (although there are some very important exceptions, such as at the federal level in the United States). Scenario analysis, as in climate change impact assessment, tends to involve simulation of the effects of different scenarios, although in water resources assessment these tend to be different demand and operational scenarios rather than different climate scenarios. Stakhiv argues that if water managers already adopt a scenario-based approach, as at the federal level in the United States, climate change therefore does not cause any additional *conceptual* challenges to water management: Climate change can be regarded simply as an extra type of scenario. However, the uncertain nature of climate change and the potential for nonlinearities in impact mean not only that the range of scenarios conventionally considered may be too narrow but also that a larger number of scenarios must be evaluated. In practice, scenario-based approaches are used in few water management agencies, and adoption of scenario analysis would challenge conventional water management practices in many countries.

Risk analysis involves assessment of the risk of certain thresholds being crossed under different possible futures. It generally involves stochastic simulation of hydrological data to develop a sampling distribution of possible futures. In principle, climate change can be incorporated into risk analysis by changing the underlying population from which data are generated according to climate change scenarios. Matalas discusses the role of stochastic simulation in the context of climate change and argues that given the wide range in futures that often is simulated by assuming a stationary climate, the operational assumption of stationarity may remain appropriate in the face of climate change in some regions. However, it is possible that climate change could generate futures outside those produced under stationarity, and it cannot be assumed that climate change can be ignored in all circumstances.

The second main issue is that of decisionmaking under uncertainty. This issue was widely investigated during the 1960s and 1970s, largely in the context of uncertainties about demands or the precise distribution of floods and droughts over the short and medium terms. Climate change has revived interest in decisionmaking under uncertainty, and several analyses of different techniques have been published. There still is considerable debate. Hobbs, for example, concludes that Bayesian approaches involving allocation of probabilities to specific o u tcomes are more suitable than

Dempster-Shafer reasoning (which requires the analyst to assign probabilities to ranges— perhaps overlapping—of outcomes), but Luo and Caselton conclude the reverse. Particularly significant is the issue of assigning probabilities to alternative possible futures. Hobbs *et al.* note unease among water planners in assigning subjective probabilities to different futures.

Planners of water resource and flood protection schemes conventionally cope with uncertainty by adding a safety factor to design estimates. This safety factor usually is defined arbitrarily.

As part of a review of water resource design practices in the UK, a more formal approach to calculation of this safety factor, or "headroom," h as been developed. Th is p rocedure identifies eight sources of supply-side uncertainty and three sources of demand-side uncertainty, each of which is given a score. The total score is summed and converted into a percentage value for the headroom allowance (with a maximum of 20%). Climate change is included as one of the supply-side uncertainties; its score depends on the range of estimates of supply-yield under four defined climate change scenarios. Although this approach has many arbitrary elements, it does represent a systematic approach to the treatment of climate change uncertainties in water resources assessment.

Different aspects of the water sector have different planning horizons and infrastructure lifetimes. The parts of the water sector with long horizons and lifetimes need to take a different approach to climate change than parts with shorter lead times; one assessment and decision methodology will not be suitable for all managers.

Factors Affecting Adaptive Capacity

From the beginning of human attempts to shape the water environment to human benefit, water management has dealt with the variability of the native supply of water and the v a riability of demands for the use of water. Great strides have been made in dealing with even extreme water regimes—particulary droughts—through interventions on the supply and demand sides. Drought management planning is playing an increasing role in many water management agencies, lowering their susceptibility to drought impacts. Thus, in some ways the prospects of a change in the resource base—perhaps characterized by lower mean supplies and higher variability—represent only a sharpening of traditional challenges to water management. There are three important differences, however. First, future climate change is highly uncertain at spatial and temporal scales that are relevant to water management: All we know is that the future may not necessarily be like the recent past. Second, as noted above, the potential pervasiveness of these changes across large regions presents challenges that preclude some traditional steps of adaptation and requires innovative approaches that go beyond experience to date. Third, climate-induced effects may be nonlinear, carrying potential for surprises beyond those inco rporated in traditional water management.

The ability to adapt to climate variability and climate change is affected by a range of institutional, technological, and cultural features at the international, national, regional, and local levels, in addition to specific dimensions of the change being experienced. Among the most important features are the following:

1) The capacity of water-related institutions, consisting of water agencies' authority to act, skilled personnel, the capability and authority to consider a wide range of alternatives (including but not limited to supply-side and demand-side interventions) in adapting to changed conditions, the capability and authority to use multi-objective planning and evaluation procedures in the assessment of policy alternatives, procedures for conflict resolution, and incentives to undertake serious *ex post* analysis of policies and projects to learn what has really worked. For example, O'Connor *et al.* found in the Susquehanna River Basin, USA, that experienced full-time water managers are more likely to consider future scenarios ir their planning than part-time managers.
2) The legal framework for water administration that always constrains, for better and for worse, the options that are open to water management. Naturally, laws change as needs change, but the changes are slow and greatly lag changing needs. In many countries, the legal framework for water management is moving toward increasing environmental protection (e.g., the European Union's habitats directive). Such a direction poses further constraints on options to address climate change, but if the move reflects an increasing concern with *su stain able* water management (however defined), opportunities for considering adaptation to climate change are increased.
3) The wealth of nations in terms of natural resources and ecosystems, human-made capital (especially in the form of water control systems), and human capital (including trained personnel) that determines what

nations can "afford to commit" to adaptation. This should include the ability and willingness to transfer wealth among population groups and regions within a country and among nations. This is the major constraint on adaptation to climate change in poorer countries.

4) The state of technology and the framework for the dissemination (or monopolization) of technology, especially in the fields of bioengineering of drought-and salt-resistant varieties of plants and techniques for the desalination of seawater.

5) Mobility of human populations to change residential and work locations in response to severe climate events or climate change. This is a major factor in coastal and island areas. Mobility is severely hampered by population pressures, especially in tropical island settings.

6) The *speed* of climate change is crucial in determining the capabilities of societies to adapt and change water management practices. Speed of change and the *cumulative extent* of change affect the impacts on society in nonlinear fashions.

7) The complexity of management arrangements also may be a factor in response. In principle, the fewer agencies involved in water management, the easier it will be to implement an adaptation strategy (although the structure within the agencies will be very important). If there are many stakeholders to involve—perhaps with c o nflicting requirements, management goals, and perceptions and each with some management control over part of the water system—it may be more difficult to adapt to changing circumstances. There is evidence that in some mature infrastructure systems, there may be substantial oportunities for increasing the resilience of water resource systems through institutional changes as well.

8) The ability of water managers to assess current resources and project future resources. This requires continuing collection of data and the ability to use scenarios with hydrological models to estimate possible future conditions.

Whether adaptation takes place or not may be heavily influenced by the occurrence of extreme events. Such events often are catalysts for change in management and may serve two roles. First, they may expose failings in the current water management system. Second, they may raise the perception among decision makers of the possibility of climate change—even if they cannot be attributed directly to climate change.

Recent experience with extreme events shows that many societies are extremely exposed to loss and damage during extreme events, especially floods. At first, it may appear that this implies that existing adaptive techniques, as widely used by water managers, are not working as expected to minimize risk and loss (some loss will always be inevitable because no flood protection scheme can provide complete protection): Adaptation is not working. However, there is extensive evidence that social vulnerability to extreme events is serious and increasing and that this exposure to hazards has been significantly increased by public and private development with insufficient regard for known hazards. In the United States there was more damage from hurricanes between 1990 and 1995 than there was between 1970 and 1990, after adjustments for inflation, even though both periods had low hurricane frequency. Changnon *et al.* analyzed the dramatic increase in dollar losses of insured property in the United States, which reached US$840 billion in the 1990–1994 period, and conclude that changes in weather and climate were not primary causes. Detailed meteorological analyses came to the same conclusion for flooding losses.

Thus, societies' failure to adapt to extreme events *in the broadest sense* (i.e., by "allowing" risk-prone development) appears to have been largely responsible for increased damages, and that failure has not improved with time. It also appears that political decisions may have produced maladaptive results. In the United States, insurance has been a leading instrument for hazard awareness and post-event recovery. After 30 years of promotion, education, and subsidized premiums, only 20% of residents in floodplains were insured by the late 1990s. These failures to take advantage of insurance suggest that even wealthy societies adapt poorly to foreseeable hazards.

The residual damages of hazard events also are inequitably distributed across populations. This was shown clearly by s tudies of Hurricane Andrew in Florida, leading the director of the Pan American Health Organization to state that "those who lost the most had the least to lose". Hurricane Mitch devastated Central America in 1998, exhibiting the extreme vulnerability of that region. Among the responsible factors were lack of land-use planning, deforestation, and inappropriate consumption and production systems.

Thus, available evidence concerning the effectiveness of adaptation to meteorological and geologic hazards indicates poor levels of individual and social adaptation to hazards. This failing extends well beyond the water management sector as conventionally defined and can be argued to reflect weaknesses in development control, planning guidance, public education, and fiscal incentives. The foregoing examples indicate that having the ability to adapt to change is not the same as actually adapting to change: The tools often are not used, for a variety of reasons.

Water managers are accustomed to adapting to changing circumstances, many of which can be regarded as analogs of future climate change, and a wide range of adaptive options has been developed. Supply-side options are more familiar to most water managers, but demand-side options increasingly are being implemented. Water management is evolving continually, and this evolution will affect the impact of climate change in practice. For reasons noted above, climate change is likely to challenge existing water management practices, especially in countries with less experience in incorporating uncertainty into water planning. The generic issue is incorporation of climate change into the types of uncertainty traditionally treated in water planning.

Integrated water resources management (IWRM) increasingly is regarded as the most effective way to manage water resources in a changing environment with competing demands. IWRM essentially involves three major components: explicit consideration of *all* potential supply-side and demand-side actions, inclusion of all stakeholders in the decision process, and continual monitoring and review of the water resources situation. IWRM is an e ffe ctive approach in the absence of climate change, and there already are many good reasons for it to be implemented. Adopting integrated water resources management will go a long way toward increasing the ability of water managers to adapt to climate change.

References

Raskin, P., P. Gleick, P. Kirshen, G. Pontius, and K. Strzepek, (1997). *Water Futures: Assessment of Long-Range Patterns and Problems. Background Report for the Comprehensive Assessment for the Freshwater Resources of the World.* Stockholm Environment Institute, Stockholm, Sweden, 78 pp.

Shiklomanov, I.A., (1998). *Asssessment of water resources and water availability in the world.* Background Report for the Comprehensive Assessment of the Freshwater Resources of the World. Stockholm Environment Institute, Stockholm, Sweden, 88 pp.

Stern, P.C. and E.W. Easterling, (1999). *Making Climate Forecasts Matter.* National Academy Press, Washington, DC.

Tobin , G.A. and B. Montz, (1997). *Natural Hazards: Explanation and Integration.* Guilford Press, New York, NY, USA

Bibliography

Aber J. (2001). Forest processes and global environmental change: The effects of individual and multiple stressors on forests. *BioScience* 51: 735–751.

Aber, J., Neilson, R.P., McNulty, S., Lenihan, J.M., Bachelet, D. and Drapek, R.J., (2001)."Forest processes and global environmental change: predicting the effects of individual and multiple stressors", *Bioscience*.

Ayres MP, Lombardero MJ. (2000). Assessing the consequences of global change for forest disturbance from herbivores and pathogens. *Science of the Total Environment* 262: 263–286.

Ayres, R U and Simonis, U E, eds (1994) *Industrial Metabolism – Restructuring for Sustainable Development*, United Nations University Press, Tokyo, Japan.

Barnett, J. (2003). Security and Climate Change. In *Global Environmental Change* 13(1):7-17.

Bengtsson, J., Nilsson, S.G., Franc, A. and Menozzi, P., (2000). "Biodiversity, disturbances, ecosystem function and management of European forests", *Forest Ecology and Management.*

Dale VH, Joyce LA,McNulty S,Neilson RP. (2000). The interplay between climate change, forests, and disturbances. *Science of the Total Environmen*t 262: 201–204.

De Soysa, I. (2005). Filthy Rich, Not Dirt Poor! How Nature Nurtures Civil Violence. In *Handbook of Global Environmental Politics,* P. Dauvergne (ed). Edward Elgar, Cheltenham.

Diehl, P.F. and Gleditsch, N.P. (eds.) (2001). *Environmental Conflict.* Westview Press, Boulder, CO.

Edwards, Paul Geoffrey; Miller, Clark A. (2001). *Changing the atmosphere: expert knowledge and environmental governance.* Cambridge, Mass: MIT Press.

Emanuel K (August 2005). "Increasing destructiveness of tropical cyclones over the past 30 years" (PDF). *Nature* 436 (7051): 686–8.

Flannigan MD, et al. (1998). Future wildfire in circumboreal forests in relation to global warming. *Journal of Vegetation Science* 9: 469–476.

Führer, E., (2000)."Forest functions, ecosystem stability and management", *Forest Ecology and Management.*

Georges, N.M. (2006). Solid Waste as an Indicator of Sustainable Development in Tortola, British Virgin Islands. In *Sustainable Development* 14:126-138.

Giddens, A. (2009) *The Politics of Climate Change.* Polity Press/

Hansen AJ,Neilson RP, Dale VH, Flather CH, Iverson LR,Currie DJ, Shafer S, Cook R, Bartlein PJ. (2001). Global change in forests: Responses of species, communities, and biomes. *BioScience* 51: 765–779.

Holling, C.S., (1973)."Resilience and stability of ecological systems", *Annual Review of Ecology and Systematics.*

Homer-Dixon, T.F. (1999). *Environment, Scarcity, and Violence.* Princeton University Press, Princeton, NJ.

Hughes, B B (1999) *International Futures: Choices in the Face of Uncertainty*, Westview, Boulder, CO.

Hulme, M. (2009) *Why We Disagree About Climate Change.* Cambridge: Cambridge University Press.

Janssen, M A (1998) *Modelling Global Change – The Art of Integrated Assessment Modelling*, Edward Elgar, Cheltenham.

Kempton, W. (1991) Public understanding of global warming. *Society and Natural Resources*, 4(4), 331-345.

Klug, J. L. and Cottingham, K. L., (2001). "Interactions among environmental drivers: community responses to changing nutrients and dissolved organic carbon", *Ecology.*

Leiserowitz, A. (2007a) *Public Perceptions, Opinion and Understanding of Climate Change – Current Paterns, Trends and Limitations.* United Nations Development Program, Human Development Report Office, Occasional Paper (no 2007/31).

Lorenzoni, I. and Pidgeon, N.F (2006) Public views on climate change: European and USA perspectives. Climatic Change, 77, 73-95.

MA (2003). *Ecosystems and Human Well-being; a framework for assessment.* Island Press, Washington, DC.

McKibben, Bill (2011). The Global Warming Reader. New York, N.Y.: OR Books.

Murdoch, W.W., (1973)."Diversity, complexity, stability and pest control", *Journal of Applied Ecology.*

Raskin, P., P. Gleick, P. Kirshen, G. Pontius, and K. Strzepek, (1997). *Water Futures: Assessment of Long-Range Patterns and Problems. Background Report for the Comprehensive Assessment for the Freshwater Resources of the World.* Stockholm Environment Institute, Stockholm, Sweden, 78 pp.

Rayner, S and Malone, E L, eds (1998) *Human Choice and Climate Change*, Battelle Press, Columbus, OH.

Rotmans, J and de Vries, B, eds (1997) *Perspectives on Global Change*, Cambridge University Press, Cambridge.

Ruddiman, W. F. (2003). "The anthropogenic greenhouse era began thousands of years ago". *Climate Change* 61 (3): 261–293.

Ruddiman, W. F., Vavrus, S. J. and Kutzbach, J. E. (2005). "A test of the overdue-glaciation hypothesis". *Quaternary Science Review* 24 (11).

Shiklomanov, I.A., (1998). *Asssessment of water resources and water availability in the world.* Background Report for the Comprehensive Assessment of the Freshwater Resources of the World. Stockholm Environment Institute, Stockholm, Sweden, 88 pp.

Schmidt, G. A., Shindel, D. T. and Harder, S. (2004). "A note of the relationship between ice core methane concentrations and insolation". *Geophys. Res. Lett.*31 (23).

Stern, P.C. and E.W. Easterling, (1999). *Making Climate Forecasts Matter*. National Academy Press, Washington, DC.

Tilman, D. and Downing, J.A., (1994). "Biodiversity and stability in grasslands", *Nature*.

Tobin , G.A. and B. Montz, (1997). *Natural Hazards: Explanation and Integration*. Guilford Press, New York, NY, USA

Wagner, Frederic H., (ed.) *Climate Change in Western North America: Evidence and Environmental Effects* (2009).

Waldrop, M (1992) *Complexity: The Emerging Science at the Edge of Order and Chaos*, Simon and Schuster, Toronto.